The Tidemarsh Guide

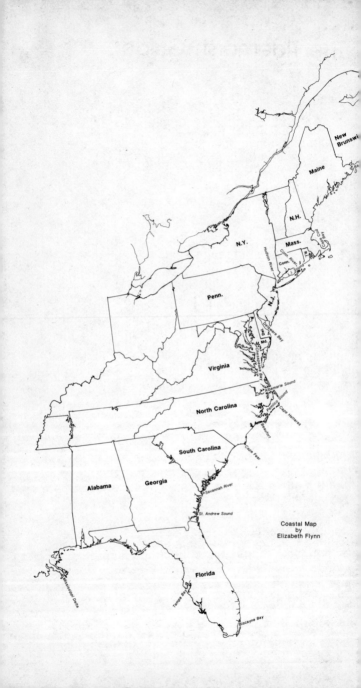

Coastal Map
by
Elizabeth Flynn

The Tidemarsh Guide

by
Mervin F. Roberts

illustrated by
Sandra G. Power

A Dutton Paperback
E.P. DUTTON, NEW YORK

Library of Congress
 Catalog Number: 79-63522

ISBN 0-525-93080-9

Published in the United States by E.P. Dutton,
a Division of Sequoia-Elsevier Publishing
Company, Inc., New York.

First Edition 1979

Manufactured in the USA by Eastern Graphics, Inc.,
Old Saybrook, Connecticut

Acknowledgements

I did get a lot of help and without it this book would not be. I cannot speak for poets but in scientific writing, I am sure no one is an island. Another thing I am sure of is that the mistakes are all mine. Wishfully, they don't weaken the fabric of the "big picture".

This "big picture" was the suggestion of my wife Edith May who advised me to take my reader out on the marshes, and this is what I hope will happen. For her thoughtfulness I rewarded her with the job of typing the manuscript from my illegible handwriting. Over the years my mother Esther Roberts has emphasized the point that English prose can be both correct and interesting. I hope I haven't disappointed her. William "Bill" Pike of Eastern Graphics agreed to share in this venture before he saw so much as a page of manuscript. He has true grit. When I thought I knew it all, John A. Seckla took me down a notch, as he taught me how to shuck an eel, Connecticut style. My original title ran on and on, complete with geographical limitations and disclaimers. Nathan Shippee helped me trim it. Dr. William A. Niering, Director of the Connecticut Arboretum republished a primer about tidemarsh plants that I wrote in 1968, and he thus encouraged me to continue writing about tidemarshes.

James N. Barnes and Dennis Costello, two local high school teachers, looked over my shoulder and rescued me from some potential pratfalls. Mrs. Martha Roberts McNair and Miss Nancy Roberts patiently read typescript and galley proofs—a tedious job at best. I'm not sure I would have stuck it out had my father been the author. Paul Spitzer and William G. Burt III, both of Old Lyme, and both experienced birders with mud-between-the-toes graciously advised me about the species which actually nest on the marshes. When the bits and pieces began to pile up Van C. Smick and Bob Pelletier sorted them out and made a book out of stacks of words and pictures. I am indebted to them for their good-natured efforts. Peter Chase and his ever-patient staff at the Old Lyme-Phoebe Griffin Noyes Library searched out books, references, and periodical reprints which I feared were lost in the dusts of time. Without their gracious help I would still be scratching.

M.F.R.

for
DR. RICHARD CIFELLI
who
got me to appreciate that there is more
about a clam than just the eating of it.

CONTENTS

THE TIDAL MARSH, GENERAL CLAIMS

When you contemplate the globe of our earth and the distribution of living things, it becomes apparent that vast tracts of land (the great deserts and the polar ice fields and the stone-covered mountains, to name a few) are inhospitable and in fact are virtually lifeless. Every commercial fisherman knows that the same is true for vast tracts of the ocean seas. Russians cross the North Atlantic to search for cod, halibut and haddock in waters off Newfoundland. Japanese cross the Pacific to search for anchovies and tuna off the coast of Chile. Admittedly, these two examples are not quite as simple as I make them out here, but they are true enough to stand up in argument, even among well-informed people. One small side issue is that there are frequently unexploited food supplies nearer home, but they are locally less desirable because of taste or custom or habit or some other human inconsistency. Mankind's diverse eating habits are on a par with the diversity of language.

A second small side issue is that as human populations increase, fishing grounds are overworked and fished out. A third small side issue is that of pollution of seafood habitat.

The side issues notwithstanding, it is a fact of life that certain natural conditions encourage the production of seafood (and food in general) and certain other natural conditions inhibit or absolutely deny such production. When a food scientist colors a map to show where productivity exceeds the requirements for maintaining a stable local population, he ends up with a shocker. Tremendous areas of land and sea are unproductive. The so-called "exporters" or "breadbaskets" are few and far between.

With these last paragraphs as a frame of reference, consider the tidal marsh. Its productivity per unit area is as high or higher than any mountainside, lake, plain, valley, bog, Philippine rice paddy or Kansas wheat field.

Many tidemarshes produce more than five tons of animal and plant life (dry weight) per year with absolutely no human effort. On the other hand, the best fertilized, pesticide treated, harrowed, cultivated, and irrigated agricultural land produces little more than the same five tons per year.

Tidemarsh meadows have produced hay crops continuously for 300 odd years on the U.S. East Coast. Hay-making commences with marking off and then recognizing the meets and bounds of the properties. Old land record books are full of detailed descriptions of the land, the rights-of-way and of course, the boat landings which were so necessary when the tidemarshes were separated from the mainland by water.

Timing the cutting and transport of hay was a tricky business because it depended not only on the season, the prevalence of biting insects, and the rain or threat of rain, but also on the tides. Horses or oxen pulled wagons of hay onto barges which were poled or towed across water. Meadows on mainland were easier to harvest, but examination of maps and deeds makes it clear that in most places the quantity of hay on the mainland did not suffice to meet the demand. The labor of hauling the hay, even over water, was justified when you consider that this hay made excellent bedding for cattle and did not subsequently contribute weed seeds to the fields when the manure and dirty bedding was spread. It is a fact that salt hay seeds just don't germinate in a non-salty environment. Of course, these saltmarsh hay fields did not need fertilizing. They were free of stones, level and they contained no thistle or poisonous vegetation; so, even though the grasses were not as succulent as timothy hay or corn ensilage, they were certainly worth the effort of harvesting so long as the greenhead flies were not around. Perhaps haying was easier on windy days when the flies and mosquitoes were less active.

Dollarwise, today muskrats, clams, blue crabs, oysters, and baits are probably the biggest direct cash crops. The indirect cash crops are, for two examples, the money spent on duckhunting and on sport fishing.

In addition to the cash crops, there are a few items from Pandora's Box. These are the diseases transmitted by saltmarsh mosquitoes and the famous liver ailment, infectious hepatitis, which is transmitted by eating mollusks which ingested human sewage which contained the disease organism.

A tidal marsh, not only highly productive, is also a special place in terms of the plants and animals found there. Many are found nowhere else, and many others common in the tide marshes are scarce or even rare elsewhere. If you have never been on one before, or if you visit in an unscientific way now and then, perhaps this book can help you. It is an informal field guide to animal and plant life of the tidal marshes of the Atlantic Coast of the U.S.A. from Georgia to Maine. With this book you will recognize typical isopods and amphipods and decapods, but not all the species. As a matter of fact there are at least forty species of isopods from Cape Hatteras, north. Detailed descriptions at the species level are available in the formal texts where frequently a whole book is devoted to a single class or family.

This book is also informal in that it is written for casual visitors to the marsh. It is weighted to favor "popular" life forms. Subjects which interest most people are covered in more detail than those which do not. For example, I don't tell you much about tidal marsh earthworms but the fiddler crabs, by contrast, received several pages of consideration because most people who visit tidemarshes do find them and find them interesting and I anticipate that you will also.

This book is also informal, as field guides go, because where there is some relationship between mankind and a particular organism—a binding tie of economics, or poison, or disease, or parasitism—which I thought was interesting, I mentioned it. Perhaps this book is best described as a brief annotated field guide for living forms of the tidal marshes. Some examples of this treatment would be the explanations for casual visitors who are confused when they see swallows around an osprey nest or two thirty-pound turtles thrashing each other in two feet of water or a person standing waist deep in water holding the gunwale of his boat and acting like he had a painful bunion.

Why did I write a book about the U.S. East Coast tidemarshes and omit Florida? The answer is tied to the life-forms; and they take a dramatic change in Georgia. From Nova Scotia to Georgia the dominant plants are species of *Spartina* grasses, and this book was deliberately written for those marshes where *Spartina alterniflora* and/or *Spartina patens* dominate the plant life.

As we go south from Georgia, the *Spartinas* are succeeded by mangroves and many of the other life forms are also replaced by species which, like the mangroves, cannot tolerate freezing and the grinding of ice. Florida mangrove tidemarshes support many birds, reptiles, fishes, arthropods and mollusks not found elsewhere and they should properly be considered separately to do them justice.

If you are a casual reader, you might wonder why there has been such a hue and cry recently over the matter of tidemarsh protection. Are these huers and criers all birdwatcher nuts? Far from it. In the opinion of this watcher of bird-watchers, they have a good reason to be concerned.

The productivity of the tidemarsh has been studied and documented; it is tremendous. The lifeforms which depend on the tidemarsh for breeding grounds and nurseries are, in many instances, found nowhere else on God's green earth. Now, here is the kicker—tidemarshes are being destroyed by pesticides, pollution, filling, mosquito ditching and dredging at a rate that is positively frightening. For example, one-half of all the tidemarshes in Connecticut were destroyed between 1920 and 1970. But enough of that, this book is not intended as a defense of marshes; it is just your guide to them, but once in a while I get carried away.

When Sandy Power made the illustrations, I told her not to draw any popular, well-known form except where another species closely resembled it. Hence, the Canada goose and the resembling brant are shown. The well-known eel, cottontail, whitetail deer, raccoon and bald eagle are not shown although they are mentioned in the text. By these criteria I admit the redwing picture is unjustified—you will have to credit it to Miss Power's artistic license. Several species of *Spartina* grass are shown and so also are several catfish and several river herring; in each case the organisms are similar and the species are frequently encountered. There is always the temptation to illustrate a book to the point where one cannot raise it off the cocktail table, and that would be self-defeating for a tidemarsh guide.

THE WATER

Tidal marshes are unique environments in which the water currents, composition and depth are primarily influenced by the tidal action of the sea. Although this is true, it is not always the whole truth. Virtually every tidal marsh also gets some water inputs from fresh sources in addition to precipitation. Without getting involved in the "chicken or the egg" controversy, one has only to look at a map of a coastline. Most (or surely many) of the most productive tidal marshes are near sources of fresh water. Sandy beaches or rocky shores make up the bulk of the remaining shoreline and they are usually found where there is no penetration of an estuary or a stream bed or bay.

The fresh water supply to the marsh is both unpredictibally variable and predictibally seasonal. The extreme case would be the mouth of a snowfed or sometimes frozen river. Here the volume flowing downstream might vary seasonally by a factor of ten and in addition, heavy rains dump unscheduled quantities of fresh water on the marshes. So, while the salinity of the sea is fairly stable between 30 and 35 parts per thousand (that is three to three and a half percent by weight of dissolved salt), the amount of fresh water that mixes with the sea water is not at all uniform or even predictable.

Not only does the salinity of water vary with the fresh water input but it varies also with the way it is mixed. Many people, duck hunters especially, have been surprised to see their dogs lapping water from tidal creeks. Now, it is an absolute and immutable fact that Labrador retrievers are not web footed any more than are Dalmatians, and furthermore their tolerance for salt water is the same as any other dogs'. So, we call or drag our thirsty Labrador away from that water; a good "Lab" pup is worth at least a hundred dollars today and surely he will sicken and vomit and perhaps die from drinking sea water. But in at least one instance the dog did not show any discomfort at all. I went through this about fifteen years ago, and in view of the fact that I had long since earned a perfectly good University Science Degree, and I knew all about salt water poisoning, I contemplated long and hard about having the stupid dog humanely destroyed before it died in agony. Then I thought about that one hundred dollars and that led me to taste what the dog tasted and I found it perfectly good to drink. The explanation is simple: As every hundred dollar dog knows, fresh water is less dense than salt water when the temperature is the same. In summertime the fresh water might well be warmer than the sea water and this will tend to make it even less dense. So, it floats on top of the sea water, and mixing is incomplete. Thus many a tidal marsh might well be covered with virtually salt-free water during certain tides.

Another example of this state of affairs concerns a dismasted and derelict vessel some 300 miles off the coast of Brazil. The survivors on board were without water and were faced with the prospect of death by thirst or by salt water poisoning. One crazed sailor opted for the sea water. He found it to be delicious and perfectly wholesome. The boat was in line with the mouth of the Amazon River, a mere 300 miles out to sea. This is not always the case. A storm may roil up the water and for a while the salinity will be the same at all depths in the creeks and adjacent waters.

Also, a prolonged windstorm from the sea will drive sea water up into the river mouths and tidal creeks and put saline waters on even higher elevations of the marsh. If the storm occurs at the time of a perigee and/or a "spring tide" when the moon and sun combine forces to raise the tide to more than average heights, then sea water, virtually undiluted, may find its way into areas generally only barely brackish. When this happens, plants which cannot tolerate salt are killed off.

Incidentally, there are a dozen good ways to determine salinity. In the classic Volhard and Mohr chemical methods, silver nitrate of precisely known concentration is titered against a precisely measured volume of unknown concentration of salty water. The resulting silver chloride, an insoluble white powder, precipitates out until the chloride is exhausted and a colored chemical indicator shows that the reaction has been completed. The amount of silver nitrate consumed is an index of the amount of salt. Other salinity measuring methods include electrical conductivity, index of refraction and specific gravity measurements. Or a known weight of water could be evaporated and the residue salts could then be weighed. An excellent method, which I recommend, is to simply taste a few drops of the water. With training, a person can get to be at least as good at determining salinity as a dog. A few bottled samples for comparison would help a novice to develop a taste for his work. Not difficult, really! Try it.

Another aspect of the tidal marsh water composition is that some marshes sit on potable supplies. A well-driller's "point" driven ten or fifteen feet into the peat will sometimes produce drinkable water even though all the surrounding creeks are positively salty or rank or both.

Some tidal marsh water is more saline than the waters surrounding. This is a summertime phenomenon and it takes place on impermeable depressions which are flooded only during the spring tides. The trapped sea water or brackish water evaporates, becoming progressively saltier and what remains depends on the number of sunny days and the amount of rain. The depressions are called pannes and plant life, if any, consists of a few species which taste astringent and grow hardly anywhere else. These "high salinity" plants include sea lavender, *Limonium caroliniarum* and jointed glasswort, *Salicornia europea.*

The water in a tidal marsh is the primary medium for the shelter of animal life and for transport of organisms, nutrients, waste products, and the people who visit the marshes. Very few terrestrial or arboreal creatures get much long term and continuous comfort from a marsh. Those best adjusted to marsh life are the coots, muskrats, railbirds, fiddler crabs, mussels, and mummychogs. This water has a property not evident in fresh water swamps, lakes, rivers or in the depths of the oceans. It moves in one direction for about six hours. Then it stops moving (slack water) for a few minutes to nearly an hour and then it moves in the opposite direction for the next six hours and so on for as long as the moon exerts its gravitational force on the rotating earth.

The currents are certainly never much more than four knots velocity; if they were, the moving water would cut the channels deeper or wider or more channels would develop. With deeper or wider or more channels the velocity would not increase. This unceasing but never racing water movement is at the heart of the whole tidal marsh system. The sessile organisms don't need tremendous holdfasts to remain in place. The waste products are certain to be removed (except from the pannes) continously except during those short periods of slack water every six hours. A constant supply of food floods into the marsh with every new tide.

A woodchuck's hole in an alfalfa field affords only security. An eagle's nest on a mountain top affords security and also a view but little more; however, a ribbed mussel in a tidal creek-side hole has all his needs taken care of by the moving water. His food supply is delivered continuously, his wastes are carried away, and even his reproductive secretions are distributed for him by tidal action, all *gratis*.

Tidemarsh life must not only tolerate the saline waters, but must also tolerate wide and rapid fluctuations in salinity. The natural history of all tidemarsh life forms is inexorably tied to their salinity limitations and requirements. Tidemarsh organisms are assured a place in the sun because they thrive under these harsh and unpredictable conditions while other forms wither away. Some saltmarsh forms even get their sexual stimulation from subtle changes in water composition. For one example, in addition to saltiness, oysters are triggered to spawn when minute quantities of ionized copper are added to the water.

"BRACKISH WATER"

Brackish is hard to define. Let's say it is saltiness that you can taste but less than seawater salinity (approximately 35 parts per thousand of dissolved solids). Brackishness in a tidal marsh is even harder to pin down to areas since it varies with time and it also varies by depth. A slow moving creek, fed by both sea and stream may have potable water on the surface and as much as fifteen parts per thousand of salinity at the bottom at a particular moment.

As you read on you may notice that I make no distinction between Tidemarsh, Tide Marsh, Tidalmarsh and Tidal Marsh, but I try to avoid "salt marsh" if only out of deference to my old Labrador retriever.

A glance at the approximate tolerance ranges of a few life forms shows that the salinity for most tidemarsh organisms spans more than ten parts per thousand (ppt.).

ANIMALS

		ppt.
Striped Bass	*Roccus saxatilis*	0-35
River Herrings	Various genera	0-35
Male Eel	*Anguilla rostrata*	0-30
Mummychog	*Fundulus heteroclitus*	0-35
Striped Killifish	*Fundulus majalis*	20-35
Virginia Oyster	*Crassotrea virginica*	7-35
Oyster Drill	*Urosalpinx cinerea*	15-28
Blue Crab	*Callinectes sapidus*	0-33
Adult Shrimp	Various genera	33-37
Juvenile Shrimp	Various genera	15-25
Red Jointed Fiddler	*Uca minax*	5-35
Sand Fiddler	*Uca pugilator*	10-70
Marsh Fiddler	*Uca pugnax*	20-70

PLANTS

Coontail	*C. demersum*	fresh water only
Wild Celery	*Vallisneria americana*	fresh water only
Pondweed	*Potamogeton crispus*	0-5
Horned-pondweed	*Zannichellia palustris*	0-5
Anachris	*Elodea canadensis*	0-10
Watermilfoil	*Myriophyllum spicatum*	0-20
Bushy pondweed	*Najas flexilis*	over 5
Pondweed	*Potamogeton perfoliatus*	5 to 25
Eelgrass	*Zostera marina*	10-35
Widgeon grass	*Ruppia maritima*	5-40

I should repeat here what I mentioned elsewhere: the tidemarsh ends (for this book) where the wild celery begins.

THE GEOGRAPHY

Generally, tidal marshes of the U.S. Atlantic Coast are level, pockmarked sometimes with depressions (pannes) and high spots (hummocks or hammocks) and cut through by creeks and man-made mosquito ditches. They are not exposed to the full force of the sea but are found facing bays or estuaries. Many others are on the protected side of sand bars and long narrow beach islands or spits. By and large, they are composed of peat which is a mixture of roots and stems of old tidemarsh plant growth with silt from the rivers and the flotsam and jetsam of the sea. The peat may well have accumulated since the end of the last Ice Age, a matter of perhaps 10,000 to 25,000 years.

Typical tidemarsh peat is as much as twenty feet deep and it will hold more than its weight in water. It has little strength in compression, but it is surprisingly hard to tear apart. The roots of plants long since dead seem to hold it together and its earthy component is frequently high in sticky clay. When it is bone dry and properly ventilated, some of it will burn. Very old European peat is commonly used as a fuel, and burning it sometimes creates an aroma reminiscent of Irish or Scotch whiskey. Although peat burns, marsh grass fires rarely if ever ignite it. In fact, these fires rarely even kill the roots of the plants. As an aside, I might mention that in the Florida Everglades, the peat does sometimes burn during grass fires, but this state of affairs was brought about by the deliberate lowering of the water table.

The great springy sponge of a tidal marsh is honeycombed and cut up with nests, tunnels, holes and undercut banks, creek beds and mosquito ditches. All over the East Coast of the U.S. tidal marshes have been constantly exploited by both lower animals and by man. People have enjoyed talking about "the forest primeval" ever since Longfellow wrote *Evangeline*, but there is probably no such thing in tide marsh literature. Tidal marshes have been mowed, burned over, ditched, dynamited, diked, and dammed time out of mind. They are not virgin primeval areas. But they do have the ability, literally and figuratively to spring back. Short of filling them or beating them to death by the wakes of high-speed motorboats they seem to last. Sometimes the sand barrier on the sea-side breaks down or shifts and the sea waves crash down directly on the peat. Wave action will then destroy a marsh eventually, but that happens only infrequently.

On the other hand, it is possible to create marshes in our lifetime. Given the right locality and some money, an "instant marsh" can be created in a few years by raising land out of the water with a dragline and a bulldozer. When it is stabilized, cultivated and nurtured to get it started it will actually support tidemarsh life. Of course the peat will not be 20 feet deep, but actually only the few top feet support the bulk of the life-forms anyway. Perhaps the "instant marsh" will not withstand the force of the sea as well as the natural product, but even so, it seems to be a good approximation of the real thing. It is not my intention here to write a how-to-build-your-own-marsh book. There is published material describing the process in detail. My point is simply that although most marshes are old, they need not be. Really ancient marshes tend to be springy; they often bounce and quiver when one walks over them. New "instant" marshes are less springy and are either hard or soft to the foot, depending on their wetness and proportion of mud or silt mixed with the underlying sand.

The ever-present honeycombing you are sure to trip over or fall into is caused by the selective decomposition of buried roots and stems, and by the concurrent drilling, digging, thrusting, wedging activity of millions of living things over the course of thousands of years. One never knows for a certainty what will appear, and from where. I remember shooting a black duck over a tidal marsh a few years ago. He coasted nearly 200 yards after I hit him, and by the time the dog got to the point where he hit the ground, there was nothing to be seen. The grass was short and could hardly hide a sparrow, but this three pound duck was gone. The dog kept nosing around a muskrat hole in the peat and eventually I got smart and investigated. Yes, the bird in its last moments actually dove so far down a muskrat hole that it took the full length of my arm to reach in and pull it out. All this on a Connecticut River tidal marsh within a half mile of the famous landmark, the Congregational Meeting House steeple in Old Lyme.

Still other holes are of smaller diameter. Fiddler crab holes are smooth and usually unbranched; some descend gently and then curve to a more steep angle. These might be a couple of feet long, (they bottom where it will be wet most of the time), and the largest will be slightly more than an inch in diameter. Still other holes are made by other species of crabs and even by mussels in the banks of creeks.

You may have noticed that sand dunes sometimes build up to the height of multistoried buildings, but tidemarshes are always level with the sea and are completely wet only at the highest tides. It is easy to state dogmatically "if they were higher, they would no longer be tidemarshes", but that skirts the history of their nature. For the last 10,000 or perhaps 25,000 years, the polar ice cap has been melting and the seas have been rising. Old tidemarshes have been constantly rising with the sea level, growing upward with flotsam, river mud, *Spartina* roots, clam shells, you name it. The interlocking mat of *Spartina alterniflora* roots is a key to the structure of the saltmarsh; and the tolerance of this plant to saline waters is the point on which that key turns. Gribbles consume the driftwood, but the roots of the grasses persist. Trees cannot tolerate the salt, nor can they gain a secure footing, so even old marshes are free of trees. Only the specialized tidemarsh grasses thrive, and so the tidemarsh of yesteryear survives virtually intact.

Compare any old map of sand bars and tidemarshes with a recent one. You will be struck as I have been by the tremendous changes in the bars and spits and sandy beaches and, by contrast, you probably will find creeks mapped a hundred or even two hundred years ago that are still meandering through peaty tidemarsh, following the same courses that were drawn by the ancient mapmakers.

THE TIDES

It is admittedly trite for me to state, but nevertheless important for all of us to remember, that one cannot say "tidemarsh" without saying "tide." The lifestyle of virtually every plant and animal on the marsh is influenced if not absolutely regulated by tidal water movement. Much of the mystery of the behavior of tidemarsh animals, and in fact all animal life, can be explained in terms of the tides. Ask any attendant in a mental institution or any nurse in a maternity ward; they will tell you that there is always more activity at the time of the full moon, and that's the same moon that gives us the word "lunacy."

For the purposes of this book we can conveniently look at two types of tides. First are the predictable tides based on the positions and movements of celestial bodies, primarily our own earth, moon and sun. We must also consider briefly the unpredictable forces like the wind, river flow, and barometric pressure changes. Sometimes these forces work against each other to reduce or eliminate a tide and at other times they reinforce each other to create unusual highs or unusual lows. Let's stack up a few facts you already know about the tides and then I will attempt to explain a few more details. To begin with, the sun crosses the same line connecting the North and South Poles (meridian of longitude) once every 24 hours. The moon's apparent motion is slower; it takes 24 hours and 51 minutes to get around once. The major tidal effect of the moon is thus scheduled by this 24 hour and 51 minute lunar day. This is the basic reason that successive high tides on the U.S. Atlantic Coast are 12 hours and 25½ minutes apart. In addition, the earth swings around the sun in an elliptical rather than circular path, and likewise the moon swings around the earth in an ellipse. Furthermore, the earth is tilted on its axis as it revolves around the sun and the plane of revolution of the moon around the earth is not around the equator nor is it in the same plane which the earth occupies on its trip around the sun.

There are many additional motions and situations which influence tides; for example, when the path of the orbit of the moon around the earth intersects the plane of the orbit of the earth around the sun a measurable effect on the tides is experienced because the three bodies are more nearly in line. This is called the nodal point and it happens only once every 18.6 years! There is no need here to mention more than a few of the hundreds of gravitational forces which influence tides. I believe it is more important to explain in general terms why, although the tides are like clockwork, the clock is the most syncopated clock imaginable.

Tide tables are available for virtually anywhere in the world to give us the predictions based on astronomical studies and experience, but this information is not all we must know. With all the astronomical data we have at our fingertips there are still many wives and mothers of duck hunters and clam diggers who often regale total strangers with long and windy stories about men who claimed to have sat one out because they missed the tide. And many a heart has failed from the effort of wallowing through mud flats which were navigable only an hour previous. Since some marshes are best visited at high tides in boats and other marshes deserve your attention at low tide on foot, you must also be a wind and barometer watcher if you cannot afford to sit one out.

You may want to go on to the plants and animals now; or if you are already stranded and sitting one out—read on, or for less detail, skip the next few pages and start again at Tide Terms.

Tides in Perhaps More Detail
Than I Should Include in This Book

Let's review the most important constituents of tide generating force derived from the sun and the moon.

I. The moon loops around the earth in the course of about a month while the earth rotates on its axis in 24 hours. If the moon stood still while the earth rotated, it would cross your meridian of longitude in the manner of the sun, once every twenty-four hours. But, in fact, it is orbiting in the same direction that the earth is rotating, and as a result of its advance an additional 0.84 hours (51 minutes) must elapse before your meridian again passes under the center of the moon. So then, the moon crosses your meridian once every 24 hours and 51 minutes. In order to more easily explain and evaluate the impact of the other tide generating forces under review here, let's call this the average moon and arbitrarily credit it with a tide-raising force of 100.0.

II. The sun crosses your meridian of longitude once every 24 hours. It exerts gravitational pull on the water facing it and thus it too contributes to the height of the tide. In terms of the average moon (value - 100.0 mentioned previously), this effect of the average sun contributes 46.6 units to the height of the average tide.

III. It will take a lunar month of 28¼ days before the moon again crosses your meridian at the same time of day it previously did. This is because of that 51 minute lag. Twenty-eight and a quarter such lags of 51 minutes each multiply out to 1440.75 minutes or 24 hours (and 45 seconds). Then for a while the sun, the moon and the earth are about as they were four weeks previous.

Leaving you and your meridian out of our considerations for the moment we discover that once in 28¼ days the Earth - Moon - Sun are arranged in that order in a practically straight line. It also means that after half that lunar month (14⅛ days) the moon has orbited half way around the earth and the three bodies are practically lined up again, but the arrangement now is Moon - Earth - Sun.

When either of these situations obtain, the gravitational forces reinforce each other and the tide springs up to higher than average levels. This is the source of the term "spring tides" or "springs" and obviously it has absolutely nothing to do with the season of the year. The tide generating force of a "spring" might be as much as 300, that is, three times the tide range caused by the average moon.

I must mention that these two alignment situations are obvious to anyone who glances at the sky and they are also shown on many calendars. When the line-up is Earth - Moon - Sun we experience a new moon (no moon) because the hemisphere of the moon which is illuminated by the sun is away from our point of view. When the line-up is Moon - Earth - Sun we experience a full moon because, with the moon on the other side of us from the sun its illuminated hemisphere is totally in our view.

Spring Tide

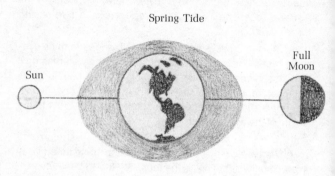

Two weeks later; again the attracting forces of moon and sun both pull in nearly the same straight line; (below) and again, spring tides result. This is the new moon.

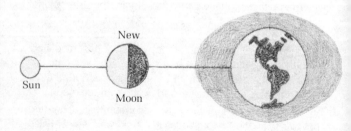

Between the full moon and the new moon, the tide attracting forces of moon and sun pull at right angles (below) and do not reinforce each other; neap tides result.

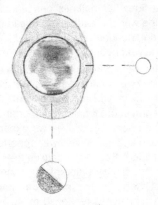

IV. Now, the circuit of the moon around the earth is not a circle, but an ellipse. The moon is nearest the earth about once a month and, when it is, we call that position of the moon "perigee." The perigee effect on the height (amplitude) of the tide is calculated at 19.1 units based on the average moon effect of 100.0. When the moon is most distant, a matter of an additional 31,000 miles from us, we call it "apogee" and the tides are commensurately weaker—that is, 38.2 units less than perigee. Perigees take place about every 27.55 days.

Now let us consider the moon's elliptical orbit and its resulting perigee and apogee with a drawing.

Moon in Apogee   Moon in Perigee

Sometimes perigee occurs when the moon is at right angles to the earth with respect to the sun (neaps). Then the sun and moon tend to work against each other in their influence on the tides. Sometimes perigee occurs when the moon is in line with the earth and the sun (springs). When there is a spring perigee, the highest and lowest astronomical tides of the year occur.

V. The path the moon takes around the earth is inclined from the plane which carries the path of the earth around the sun. This inclination (or declination) amounts to only about five degrees, but it has two important effects. First, it reduces the number of eclipses we experience because the earth, moon and sun have fewer opportunities for being absolutely aligned. Second, it causes the moon to pull on the water from a slightly different angle than the sun pulls on it most of the time. The maximum amplitude of tides caused by declination is reckoned at 12.7 units based on the tidal effect of the average moon at 100.0.

VI. When the moon or sun overhead exerts a gravitational force on the water we also experience a second effect of nearly the same magnitude. This takes place on the opposite side of the earth where there is a second high tide because the pull on the water on the far side is less than the pull on the earth. Another way to look at it is that the earth is drawn toward the sun (or the moon) more strongly than is the water on the far side. This is what causes semi-diurnal tides, that is—twice daily.

VII. There are also several forces which are primarily "once daily" (diurnal) rather than twice daily (semi-diurnal) and these are of concern only to astronomers and their like, but I believe by now I have about made my point about the "syncopated clock."

Yes, as with virtually everything in Mother Nature's domain, the tides do follow a pattern, there *is* a rhythm. As I believe I have amply demonstrated, to anticipate the nuances of this rhythm (disregarding rain, run-off, atmospheric pressure and winds) takes the efforts of astronomers. The tide tables, mentioned elsewhere, are a matter of public record, but their story is not the entire story. Some of the other inputs are a lot less predictable, but on occasion, much more profound. Many a "bay man" has remarked that "the tide was coming in but the wind blew all the water out."

The salt water flushing which is administered to the marshes by these highest tides certainly has had its impact on the question of land ownership. It also has a profound effect on the plant life. These extra high waters "weed out" virtually everything except the tried-and-true tidal marsh plants. There is no clover or poison ivy or perennial rye grass on a genuine tidal salt marsh.

On the high seas the amplitude of tides is smaller, and the tidal currents are ignored. the heavenly bodies still most strongly attract the water which is directly in line with them, and gravity still pulls the earth away from the water on the "away" side; but no great quantity moves any great distance on the high seas and the range of tides on tiny islands in great oceans is usually less than a foot. Near the great land masses with shores than run generally north-south, the water that followed the moon or the sun hits the obstruction of shallow water, reefs and sloping shores. With no place to go, it tends to pile up and then when the moon or sun has passed and no longer causes the strong pulls on this piled-up water, it rushes away in an outgoing (or ebbing) tide until the next time around.

Although the height of a tide is a small thing on the high seas, in bays and estuaries it can be profound. The classic cases, such as the Bay of Fundy, are awesome in their movement of water. Schemes are always under way to harness this tremendous source of energy.

One important reason that tides are not still higher is that the viscosity of the water reduces its ability to respond. It should be easy to imagine how small the tides would be if the seas were molasses. Even the "solid" earth itself is a viscous mass and it too bulges from the gravitational forces of the moon and sun. Sometimes it bulges as much as ten inches in diameter from low to high tide! You need not worry about it; you didn't worry before I brought it to your attention, don't worry now.

The study of tides has a history of its own, and a rich language to describe the rates of change, the duration of slack water and all the other variables. Every aspect of tide is variable and even though you just read and hopefully understood how it is that high tides on the U.S. Atlantic Coast follow each other every 12 hours and 25½ minutes, I am sorry to have to tell you that there are tides in the China Sea which often occur only once a day. This book is not the place for that explanation.

Lawyers, realtors, surveyors, and mariners are continually challenged by the matter of the high-water mark, or some other water mark. The problem comes up constantly in seashore places because many old deeds and land grants extended to a line marked by a tide. If the elevation of the land should rise or fall, the property line would shift. The business of land grants and the sovereign power of kings and other potentates to keep or relinquish rights to tidewater lands is with us all the time. In some instances in the northeastern U.S. a tract of land was deeded, surveyed and taxes were paid on it for 300 years before someone came before a judge and claimed that the property is really in the public domain. Often the property is a tidal marsh. Some old English law reserves for the crown all lands that are wet by tidal waters, or mean sea level, or mean high water, or perigee high water. When land wasn't worth much, no one took much trouble to define the terms. Today these interpretations mean millions of dollars. For example, the three big airports serving New York: La Guardia, Kennedy and Newark are *all* built on lands which were, within the memory of this writer, tidal marshes. As a matter of fact, as a boy, I netted blue crabs and dug steamer clams out of a creek only a few hundred feet from the control tower at New York International (Kennedy) Airport.

I mentioned only a few, but astronomers tell us that there are between 60 and 70 motions or relationships of celestial bodies which account for 99% of the total astronomical tidal effect. Although the number of parameters which control the remaining 1% of the astronomical tide effect is mind boggling, it is not meaningful in terms of inches of water when you consider local conditions. For example, the barometric pressure caused by weather has much more influence than most of those minor celestial forces. A one inch drop in the mercury barometer will immediately cause a one foot rise in tide. The converse is also true. Also, as the tide changes, it influences the barometric pressure. So then a barometer at or near the sea will follow the atmospheric pressure created by the weather, and it will also follow the astronomical tides caused mostly by the sun and the moon.

So, to wrap it up and add a few more parameters: tides on the marshes are the result of: the combined effects of gravitational pull by celestial bodies on each other; the centrifugal effects of rotation; the push or pull of atmospheric pressure as displayed on a barometer; the force, direction and duration of the wind; the input of fresh surface water flowing into a marsh from uplands; the effect of ice which may trap what would otherwise be free-flowing waters; the viscosity of water especially in shallow estuaries and long basins like Chesapeake Bay and the Hudson River; and even the gravitational attraction of the earth toward the moon which pulls the earth away from the water which is on the far side of the earth, and creates that second high tide about twelve hours and twenty-five and a half minutes after the previous high tide.

Don't take the dimensions of these tide diagrams literally. They are deliberate gross exaggerations. To begin with, the diameter of the earth is nearly 8,000 miles (about 48,000,000 feet) and the tidal effect is always less than fifty feet - in most places only a few feet. If the earth and the tide were drawn to the same scale the tidal effect just would not show, it would be lost in the thickness of the lines. Also the spacing of the earth, moon and sun were deliberately drawn to demonstrate their effects but not their real distances. The moon averages a quarter of a million miles distance from the earth and the sun ranges in the neighborhood of ninety-three million. So if we drew the earth and moon one inch apart, then the sun would properly hang out (4 x 93 ÷ 12) thirty-one feet away. The relative sizes are also wrong. Earth is about 8,000 miles, moon is a bit over 2,000 and sun is estimated at 864,000 miles. If we draw earth with a diameter of one inch, the moon should appear as a one-quarter inch disc and the sun would have to be one-hundred and eight inches (nine feet) in diameter.

TIDE TERMS

TIDE - The vertical movement of water.

TIDAL CURRENT - The horizontal movement of water.

SPRING TIDE - Nothing to do with the season of the year. A spring tide occurs when the moon, sun and earth are aligned, once every 14⅛ days. This tide "springs up" higher than the usual high and lower than the usual low because the gravitational forces of sun and moon combine in a straight line. It happens once when the moon is between the sun and earth, and again when the moon is on the far side of the earth. Spring tides coincide with the new moon and the full moon. Perigees (27.55 days) and springs (two every 28¼ days) are slightly out of time with each other so they happen together only infrequently.

NEAP TIDE - When the moon is 90 degrees out of line with earth and sun, its effect and that of the sun do not reinforce each other, and the high tides are lower than average high and low tides are higher than average low. This happens one week after a spring tide; that is, about twice monthly. Neap tides coincide with the first and third quarters of the moon.

PERIGEE - The moon is nearest the earth in its elliptical orbit. This makes for a higher tide by about 20% over an ordinary high tide and 20% under an ordinary low. Since the center of the orbit and the center of the earth are not the same, this near position happens only once in 27.55 days.

APOGEE - The moon is at its greatest distance from earth and has its least influence on the tide. High tides are less high. Low tides are less low. Apogees occur approximately two weeks after every perigee.

RANGE - The distance vertically between high and low tides. It varies from place to place. On the high seas it is hard to measure. Some tiny islands in great ocean basins experience very small ranges of perhaps only one foot; on the shores of some continents where the water is shallow and mouths of estuaries are narrow, two tides of over forty feet are daily events.

INTERVAL - The time between successive similar tides, called the interval, varies tremendously. The Atlantic Coast of the U.S. generally follows the 12 hour and 25½ minute pattern between successive tides, but the others don't. Some other parts of the world have 24 hour intervals between high tides; and in Tahiti I am told that high water always occurs at noon.

EBB- The falling tide, the outgoing tide.

FLOOD - The rising tide, the incoming tide.

ESTABLISHMENT OR TIDAL ESTABLISHMENT - The time between the passing of the moon and the top of the tide at that locality. The establishment may in some places lag the moon by as much as eleven hours. This lag is due, mostly, to the drag caused by the viscosity of the water.

SYZYGIES - These are the full and new moons which cause the spring tides. These springs are the times of the highest high and lowest low that can be expected on a regular monthly basis. This term was added for readers who think they know it all. It should also prove useful from time to time in crossword puzzles.

QUADRATURES - These are the times of the first and last quarters of the moon and the range is small because the moon and sun do not reinforce each other's pull. This is the time of neap tides with lower highs and higher lows than average.

AMPLITUDE - One half of the displacement between maximum and minimum; that is, half the range of a tide. The word amplitude is not used on the marshes but is part of the language of the students of the tides.

SLACK - "Slack water", the tide stands still after it has gone out to "dead low" or come in to "dead high." In river estuaries, slack water occurs when the new incoming tide just balances against the current of the everflowing river. In that instance, the slack water might take place an hour or even several hours after the time of dead low.

DIURNAL INEQUALITY - the two successive highs on the U.S. East Coast are usually not of equal height for any particular place. We call this difference the diurnal inequality of the tides. The cause is astronomical in nature and it depends on declination. The moon does not revolve around the earth directly over the equator or even in the same tilted plane which the earth occupies in its path around the sun (ecliptic), rather it takes its own trip on its own route. This apparent route is tilted (declined) from the path of the ecliptic of the earth by about five degrees.

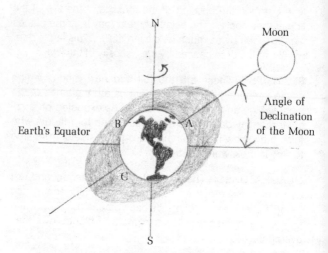

Look at this simplified but exaggerated drawing and note that when you are at point "A" and the moon just happens to be overhead, the tide is at its highest. The complimentary high tide is at point "C". Twelve hours and twenty-five and a half minutes later the earth has rotated you to position "B" for your second high tide of the lunar day, but as you can see, it is less high.

I had to mention this rather complicated celestial state of affairs because later on you will encounter a most common little fish (the mummychog) which is reported to utilize diurnal inequality in its spawning procedure.

THE VISITOR

Getting around on a marsh can be frustrating or even frightening to a novice. Frequently, after hours of back-breaking slogging through an almost impenetrable morass of *Spartina* and muck, the first-time visitor arrives exhausted at his goal, covered with bug bites and grass scratches and mud only to discover some "old time marshrat" cool and clean and relaxed tonging clams or emptying and re-baiting his killie traps and eel pots or crabbing or just laughing at the mess you got yourself into.

The difference was in attention to details which taken singly are almost absurdly simple. The novice was probably dressed (or undressed) for the beach. This garb of exposed skin is all wrong on a tidemarsh. Until you know better from experience, believe me—wear cotton tightly woven hard fabrics in summer. Washable wool is a favored fabric in wintertime. Avoid synthetic yarns, they tend to keep you too hot in hot weather, too cold in cold weather. Wear a hat. Wear tennis shoes without socks, or wear hip boots. There is no sense using anything on your legs above your ankles unless you go all the way up with protective clothing. I'll have more to say about this later on.

Choose a hat with a wide brim or a long beak. Shade your eyes with the hat; and if possible avoid sun glasses. They tend to get knocked off if you are working and sweat builds up under them. Also, too frequently you will want to wipe bugs or mud off your face, and glasses, especially large ones, will be in the way. Duck hunters avoid sun glasses because the reflections scare off the birds.

Use an insect repellant. Use it liberally. Carry it with you. It is surprising how often one gets bitten on a tidemarsh a month before or a month after the insect season on the upland.

Tell someone (someone you trust) exactly where you are going on your first trip, and don't go anywhere else. If you are planning to be gone long, take a few pieces of fruit, a couple of oranges perhaps. Oranges are good because the peel protects the fruit against bruising and the juice is a great thirst quencher and energy source.

Avoid carrying gear in both hands. Try to keep at least one hand free, and remember to carry a light oar for feeling the depth of the mud. Be familiar with landmarks. Carry a watch, a tide table, and a map. Mark your route on the map. A small CB radio is probably a good investment if you have someone on the same channel monitoring you. Transmission over marshes is usually very good because there are few or no obstructions to the straight-line radio waves.

If you are using a boat, be sure you know how to operate it before you leave the dock. If there is even a remote possibility that night will fall before you return, bring a flashlight. If this is your first experience in a tidemarsh, especially if you are in a strange place, hire a guide. This is especially important if you plan to collect any specimens, since the applicable laws may be confusing and a guide would likely be knowledgeable and licensed for whatever is legal.

Obey any "No Trespassing" signs; it will amaze and embarrass you to discover that you are being watched by more eyes than just those of rails and fiddlers.

You don't have to be web-footed, but it probably would help. Let's look at what we are getting into. Most creeks have soft mud bottoms. The depth of mud can frighten a person if he enters unwarned, but very few actual drownings are recorded. The mud does provide a little buoyancy and there is generally a hard bottom within two feet. This is the mud ("clam mud") in which clamdiggers "tread" for hardshell littlenecks, cherrystones and chowder clams. Let's consider how a clamdigger capitalizes on the morass that frightens so many uninitiated people. Treading is a summer-time occupation and it is accomplished when the tide is deep enough to pole or scull or row a boat into a mud-bottomed creek and shallow enough so the treader with his shins half in the mud can at least keep his nose above the water. To tread for clams one must assume this position and then with a sieve on a long stick, or a steady foot available, a treading movement of both bare feet in the water, with both hands on the gunwale of a wide boat will produce clams. When clams are felt, underfoot, they are raised on the top of one foot and taken in hand or into a small sieve or net for a quick rinse before being deposited in the boat. Under ideal conditions a person can tread a bushel in a tide, or before he turns blue or wrinkled. The mud these clams are in is too soft to support a treader's weight, but stiff enough to support the clams. At dead low tide the mud may be exposed and impossible to negotiate or there may still be several feet of water in the creek. So then, this is the consistency of a typical tidal creek bottom. Near the entrance to a bay or other larger body of water, the bottom frequently hardens up with packed sand. Low velocity meandering creeks tend to have more depth of softer mud than do straighter, faster channels.

The shores of a marsh are sometimes mudflat and sometimes submerged peat and sometimes even sand. The best way to find out is with an oar or a crabnet handle. An oar is better since it is less liable to break when it is being pulled out of the mud and with an oar, the handle end remains clean. If you are not familiar with a marsh, it would be a good idea to carry an oar as you walk, even over the upland. A few other good ideas learned by the author in the course of about fifty years are, in no particular order:

Always wear footgear on strange marshes. Some have been dosed with broken beverage bottles, and in soft mud it is impossible to predict what one is stepping onto. If the temperature does not demand wool stockings under boots, then at least wear canvas-top tennis slippers, perhaps best without socks. Most of the mud can be rinsed or scrubbed off and the slippers will dry in the sun before your next visit. Caked marsh mud generally breaks free from fabric when it is bone dry.

Boots should be as light and loose as possible. Steel toed industrial boots are worse than useless because they are unnecessarily heavy and rigid. If you are about to buy a pair of boots, get duck hunter's or surf caster's hip boots *without* insulation. Wear only wool socks inside the boots. Choose a boot style which does not form-fit your ankle or calf. Learn to fold the boots down to the knee.

Clip off the knee clamping web or other gadgets from the inside of the boot. In the surf or in snow these hitches are useful but in clam mud these devices are dangerous. Once, just once, in your lifetime you will want to get out and leave one or two boots behind. The belt straps on your boots are useful, and your trousers should be furnished with a reliable belt, but generally the boots are worn folded to the knee and are drawn up only when fording creeks or working with a boat.

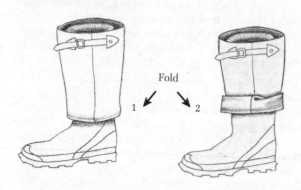

Fold

1   2

Wear long trousers when "exploring." Some reed grasses, dune grasses and rushes are spiny and/or sharp-edged. The long trousers and long-sleeved shirts are also good insurance during insect season. A good insect repellent in warm weather goes without saying. If the hair on the top of your head is thin, a hat is also a good insurance policy, both for insects and for the sun.

Now, as to the boat, there are many opinions. Most strong opinions are from people with special interests. Here are a few:

The *canoe* is fast and quiet and it draws very little water. Metal canoes are vulnerable to puncture but are light. Square-end canoes are hermaphrodites which permit small outboard motors without outrigger gadgets. The canoe is tippy and requires disciplined or trained passengers.

Round bottomed row boats—"outboards" of the Thompson-Penn Yan-Lapstrake-Lyman style—are a compromise between a flat-bottomed scow and the canoe. They are light, easy to row, scull, pole, motor, or even paddle. Nowadays, these wooden boats are expensive, but anyone who ever had one will look back at it with fond memories. Your name should be branded into the keel, and such a boat deserves to be furnished with a combination lock and a length of stout chain.

Fiberglass impregnated with resins makes a rugged hull, easy to clean and inexpensive by comparison with the better wood boats. The better fiberglass boats tend to be heavy. If you plan to haul your boat over the meadows or push through soupy water, you might be happier with a light, inexpensive aluminum hull and thereby sacrifice the quiet springiness of wood or the ruggedness of fiberglass for the savings in weight.

Wooden scows make great marshmen's boats. One small person needs nine or ten feet, would do better with eleven or twelve. To take a guest, say twelve feet minimum and a guest with a dog or child call for thirteen or fourteen feet. Over fourteen feet gets unwieldy and, as you gain capacity, convenience and comfort, you lose maneuverability in shallow and narrow and winding creeks. Wooden scows are generally made with ¼ or ⅜ inch marine plywood bottoms and ¾ inch cedar ends and sides. From the gunwale to the deck, twelve to fourteen inches is appropriate and the bow is sometimes slightly pointed and sloped to streamline it for slow speed progress. When you drive a scow at much more than rowing speeds it will plane and then the shape of the bow is absolutely meaningless in calm tidal creek water. The stern is square to the bottom and is reinforced for a small outboard engine or a battery powered electric motor-driven screw. The stern should be notched or equipped with a rowlock for sculling and any of these boats should be furnished with a pair of oak or ash oars, and, for safety, a spare. Pine oars are lighter but less strong. You pay your money and take your choice. The scow should be wide enough to assure stability. You will quickly learn to walk (not teeter) around in a twelve-foot scow, something which only an acrobat could accomplish in a canoe.

Aluminum scows are light and inexpensive. You will get good value from the larger mail-order houses like Sears Roebuck, Montgomery Ward, and Herters.

Motor drives, if you find you need power, should be the smallest and lightest you are willing to settle for in this age of speed. Two horse-power is ample for marsh creek travel even on a twelve-foot boat. One point is that frequently you will want to remove the motor from the transom and its weight will be a factor. Another point is that high speed through tidal creeks is not usually productive of anything, and really high speeds may destroy a lot of life and tear down banks or silt up the creeks. A final point is that most problems with outboards in marshes occur when the water pumps fail. Some low horsepower motors (and all electrics) are air cooled and this may be an important consideration for you.

Accessories for your boat depend mostly on whether it is motorized. State and/or federal regulations for motorboats call for life jackets, and in some cases even a fire extinguisher and a whistle may be required. An anchor, attached to at least fifty feet of anchor line is good insurance. Never leave the anchor in the boat when you get out in the marshes. There is always one stiff puff of a breeze or one swell at the top of the tide to carry your boat away. Tie or clip the anchor line at both ends, that is, don't splice it into the boat or onto the anchor. You may need it for a tow or for tying up. A bailer is useful; so also are a few pails. Plastic pails are rustproof and quiet.

Boats are positioned in mud-bottomed marsh creeks sometimes with poles. In fact, Herter's offers a boat with a hole in a well at either end through which a pole is driven into the mud. This moors the boat, but permits it to move vertically with the tides. Some duck hunters use a centerboard boat, not for sailing stability but for mud-flat mooring. Of course the anchor is the most common device and the most common anchors are the ones shaped like umbrellas or mushrooms. Actually, a ten pound piece of railroad iron suffices in most tidal creeks if you leave your boat in the water. Don't leave a wooden boat too long, or the gribbles will eat it up.

PLANTS

Plants with roots are convenient benchmarks for salinity, soil type and moisture. The competition for a toehold is fierce; seeds are broadcast by the thousands and runners push in all directions. During the growing season the entire upland marsh is green, but no one plant is found everywhere. Although the tidemarsh is fertile and sun-drenched, the wetness and salinity place severe limitations on plant life. There are, for example, dune grasses and peas that tolerate salt but never grow on salt marsh peat although they are abundant or dominant on sand dunes immediately adjacent to the marsh. And on the sloping marsh itself the plants succeed each other according to rules set by Mother Nature. Their seeds are everywhere and their root stocks are pushing relentlessly, but the lines are firmly drawn and there is just no *Juncus gerardi* in the mosquito ditch and there is no *Spartina alterniflora* on the higher meadows. Agatha Christie could have written a good murder plot for M. Poirot where the victim had a creek-bottom "clam mud" on his shoes but the prime suspect had only *Juncus* flowers from the upper marsh in his trousers cuff.

Some tidemarshes from Maine to Georgia have been mowed for three hundred years or more and still don't seem much the worse for wear. The grasses propagate by either seed or by root stocks and runners, and so mowing even before they go to seed doesn't seem to set them back. The harvesting techniques and tools are part of our agricultural history, but today the amount of salt hay cut is hardly more than a token of the total that grows and is eaten or decomposes to feed the plankton and the detritus eaters and then the fish, *et cetera, et cetera, et cetera.* Salt marsh hay is good barn and stable bedding and when it goes out with the manure it doesn't seed a field with the wrong crop since salt hay seed will not germinate without soaking first in salty water. As forage, it leaves something to be desired. It is nourishing and cattle will eat it, but given the choice, I understand they prefer clover, alfalfa, timothy and corn.

Draw an arbitrary line a foot or so above the mean high-water mark and there we find a band of shrubby bushes or bushy shrubs. Above them we are out of the tidemarsh with cedar trees and junipers and poison ivy and honeysuckle and bullbriars. Below those hightide bushes are the sedges and common reed grass and marsh goldenrod. Still closer to the tidal water are the blackgrass and the cattails and the *Spartinas* which are the true criterion for this book. The actual walking distance from the upland shrubs to the stem-wet *Spartinas* might be as close as fifty feet or as far as a mile depending on the steepness of the slope and on that slope you can nearly measure the elevation by the species of plants.

As you look at the pictures bear in mind that plants do vary a lot in their appearance depending on the season and on local conditions. Look at the familiar oak or maple in an open field. The tree spreads out into a well rounded globe or pyramid and the lower limbs are stout and sometimes nearly horizontal. When that same species of tree grows up in a new forest, the trunk goes straight up and there are hardly any living branches below the tight cluster at the top. The same things happen on tidemarshes. Crowding, food supply and salinity do affect the shape of plants, so be tolerant and imaginative as you compare the pictures with the living material.

This high water mark, where the shrubby bushes grow, deliniates the border of *any* marsh influenced by ocean tides. Interestingly enough, it does not deliniate the border of what most scientists call a saltmarsh. I realize that this is a bit sticky but the saltmarsh is not only wet by waters influenced by ocean tides but those waters must be salty or brackish at least from time to time to the extent that those plants which don't tolerate salt, don't grow there. Remember that the Hudson River is tidal to Albany (150 miles) but is hardly salty above Peekskill (50 miles); and the Connecticut River is tidal to Hartford (40 miles) but never salty above the bridge at Haddam (less than 20 miles).

So now, please, as you identify the plants draw another arbitrary line where the water hardly ever tastes the least bit salty. This line is more difficult for us to establish by scientific analysis but plants make the discrimination all the time. For example, eelgrass *(Zostera)* is found in salty marsh waters but a look-alike wild celery *(Vallisneria)* is limited to "fresh" water. Another example is the cattails which I discuss in more detail later on.

The cut-off for tidemarsh salinity, as I define it for this book is: Where eelgrass ends and wild celery commences.

Where narrow-leaf cattail ends and broad-leaf cattail commences.

Where the *Spartinas* end and wildrice commences.

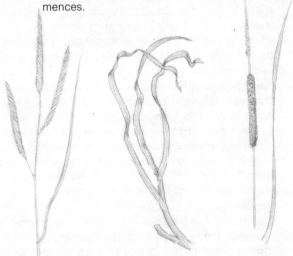

Spartina   Eelgrass   Narrowleaf Cattail

A tidemarsh, with its tangle of roots and *Spartina* stems is a natural trap and catchall for dead and dying plants and animals. One reason there is so much death and also so much specialized life in this brackish environment is that saltwater forms which drift inshore with winds and tidal currents are sometimes unable to adjust to the lower salinity of most tidemarshes, especially those in estuaries. Additionally, fresh water forms drifting downstream in rivers and creeks reach the same brackish water and these too do not always adjust to the salinity level. So they all die and then the scavenging amphipods and isopods and decapods make mincemeat of them, only to be eaten in turn by the predators such as the striped bass, the marsh wren and man.

Common names of some plants include the words *Typha*, grass, reed, rush, sedge. Unfortunately not too much sense can be made of these designations since the common names frequently disregard scientific botany.

True grasses have two rows (or ranks) of leaves on a stem. These leaves erupt from thickenings called nodes, alternately from opposite sides of the stem. Grasses have cylindrical stems, sometimes solid, sometimes hollow. Bamboo is a "woody" grass. Panic grass is a typical grass and so is rice. The common reed grass phragmites is also a grass although the leaves are not arranged in two straight rows. Let's say, at least for the tidemarshes, that a reed is simply a tall grass.

For convenience here let's separate the sedges from the rushes, although some botany texts tend to combine them. Sedges have solid stems, triangular in cross-section. Sometimes these triangles are rounded to the extent that you may not recognize them as such even when you are looking at them, but the triangle governs the pattern of leaf arrangement and so in sedges we find not two, but three ranks or rows of leaves erupting from the stem.

Rushes have hollow or pith-filled stems and sometimes hollow leaves. Flowers develop on the ends of the stems. The *Juncus* plants are rushes. Rush leaves are not arranged in ranks of two (grass) or three (sedge). Some rushes have all their leaves erupting from the area just above the root stock and others have leaves coming out of the main stem at various heights, but the rigid pattern of ranks is not apparent.

Cattails are in a family unto themselves, call them *Typha* or *Typhaceae*. They differ from grasses, rushes and sedges in that their flower stems bear the two sexes as separate compact parts on the single stem. The upper part is the male flower and the lower part is the female flower (which eventually produces the seeds.) Thick root stocks contain large quantities of edible starch.

In 1968 I wrote a booklet about tidemarsh plantlife for the Conservation Commission in my home town; and several years later Dr. William Niering, Director of the Connecticut Arboretum at Connecticut College made it a part of their reprint series. Here I draw freely from it.

MARSH ELDER

The bushes at the highest level wet by saline tides are usually marsh elders, also called the hightide bush or the highwater shrub. This plant is technically a sunflower, *Iva frutescens*, and is hard to push through but not thorny. Stems are two to five feet high and the leaves are as much as four inches long. Some leaves have coarse teeth and all are fine-haired or hairless. The greenish white flowers are small and inconspicuous. This plant is not woody like a privet or an evergreen nor is it reedy like a grass but rather somewhat in-between.

Elder

SALTMARSH ASTER

This plant is, like the elder, a technical member of the sunflower family. It grows from six to eighteen inches and is often found in low spots on the meadow of the marsh. *Aster tenuifolius* is a late bloomer and it resembles our familiar upland asters. The flowers are white or pale purple and it is found in small clusters from New Hampshire to Florida.

Aster

GOLDENRODS

Any tidemarsh from New Jersey north is likely to be peppered with these specialized species of goldenrod. Superficially they look like ordinary upland weed-patch-and-roadside goldenrods. The big difference is that these species tolerate soaking in brackish water.

Seaside goldenrod grows to a height of three feet and is crowned with a large yellow flower cluster, it is certainly spectacular when it is blooming. The leaves are long, fleshy and smooth. Technically this plant is, like the aster and the elder, a sunflower. Although some texts place it typically on the upper parts of the marsh, I have frequently seen it with stems wet daily by tidewaters. The lower leaves are wrapped around the stem. This goldenrod is *Solidago sempervirens.*

Another related plant is the slender-leaved goldenrod *Solidago tenuifolia*. This latter species is found from Virginia, north, and it grows to a height of about 30 inches. The name "slender leaf" is the best fieldmark once you recognize it as a typical goldenrod.

Goldenrod

MALLOW

The big bright pink flower that looks from a distance like a child's plastic toy—it really seems out of place on a brackish tidemarsh—this is surely the swamp rose mallow, or marsh mallow, *Hibiscus palustris*. The plant grows to a height of seven feet and has a showy display in late summer of large pink or purplish flowers. It is common between Massachusetts and North Carolina.

Mallow

WILDRICE

This is the stuff, *Zizania aquatica*, that costs upwards of seven dollars a pound when it is harvested from canoes in Minnesota. In Connecticut ducks eat it "for free." It grows to a height of ten or twelve feet and the flower cluster might be two feet of that. The lower branches of the flower cluster carry the male spikelets and on the upper branches you will find the long brown grains of rice. Rice is not really a brackish water plant; you will rarely find it in waters with salinity levels that you can barely taste. Maybe I'm wrong, maybe it just can't compete for a root-hold with *Spartinas* and so it only occurs in low salinity areas since *Spartina* seeds need to be soaked in salty water before they will germinate. Wildrice (also spelled wild rice) tastes delicious to man and beast alike.

PLANTS WITH ARROW OR LANCE SHAPED LEAVES

There are many species of wetland plants which have toothless leaves shaped like arrowheads (or wider) and whose solitary flowers rise from the rootstocks on individual stems. Their names include arrow-arum, American lotus, spatterdock, pickerelweed, several arrowheads, duck potato, bull tongue, water plantain, burhead, sea lavender, saggitaria, golden club, frogbit; and two floating plants—water hyacinth and water lettuce.

These plants are found at the borderline of the brackish or salty tidemarsh. The fiddler crab rarely burrows among their roots. Where does one draw the line? In nature there is no black and white but rather an infinite number of shades of gray. In this book I try to stop at the place where you cannot taste the salt but since salinity is constantly changing with winds and tides and surface run-off and river flows it is not a certain thing.

TAPE LEAFED WATER PLANTS

It is worth mentioning that while adult saggitaria is out of the water, young saggitaria leaves are totally under water and at that stage they are tape-like and stemless. You will find them in fresh water marshes and aquariums along with wild celery *Vallisneria spiralis*.

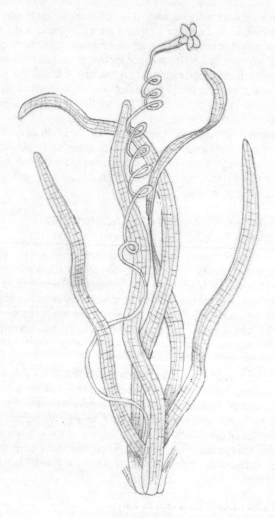

Wild Celery

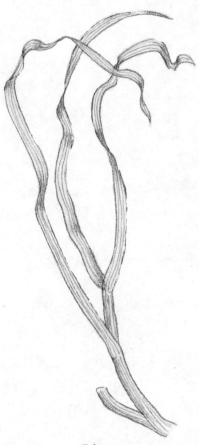

Eelgrass

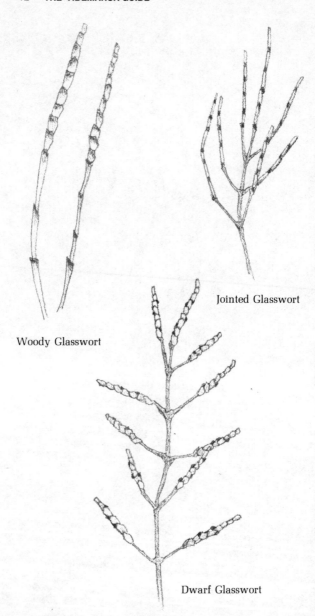

Jointed Glasswort

Woody Glasswort

Dwarf Glasswort

Nearer the sea water the tape-like water plant is eelgrass *Zostera marina*. This is the plant which provided food for brant and which was blighted nearly to extinction in 1930-1932. *Vallisneria* and *Zostera* resemble each other very closely but for their habitat.

I mention saggitaria and vallisneria here because they are both found on the margins of brackish tidemarshes. They do not live in salty water but they do tolerate occasional inundations of diluted seawater.

GOOSEFOOT FAMILY

These plants are not leafy but rather "fleshy." They are related to beets according to the botanists, but for casual tidemarsh visitors the standout feature is the lack of flat leaves. Flowers are tiny and inconspicuous. The genus we are interested in is *Salicornia* and there are several species all called glassworts. I think of it as a tidemarsh weed since it often appears in bare spots where *Spartina* and *Juncus* fail to take hold. The branches are eaten by waterfowl and humans as well. To me the taste is a mild-sour-salt. In autumn much goosefoot turns red.

The species you probably will find are:

Salicornia europea - jointed glasswort, samphire, saltwort. Main and branching stems twenty inches high. From Georgia, north.

Salicornia virginica - woody glasswort. To twelve inches high. Single stems from root to flower. From South Carolina to New Hampshire.

Salicornia bigelovii - dwarf glasswort. To twelve inches high. Main stem with flowered branches. From South Carolina to Maine.

RUSHES

Grasses have relatively slender stems and reeds have round hollow stems and rushes have triangular stems. Rushes also have a claim to fame since it was in a clump of bulrushes that baby Moses was hidden from the Pharoah. Rushes are properly in the family of sedges and they have not only the typical triangular stem but the tight flower and seed clusters to help you recognize them.

Saltmarsh Bullrush

Sea Lettuce

The saltmarsh bulrush *Scirpus robustus* grows two to five feet high and some of the leaves grow higher than the stalk. Birds eat the seeds; muskrats and geese eat the stems and rootstocks.

A shorter sedge is the swordgrass, three square or chairmakers' rush *Scirpus americanus*.

SEA LETTUCE, GREEN SEAWEEDS

These are the creek and bay-bottom leafy rootless thin green wrinkled sheets which spread to three feet in diameter over mud flats. Technically, the most common is *Ulva lactuca*, and it is seaweed—a nonflowering, seedless algae. There are several species, all similar in general appearance, all usually called sea lettuce.

Whereas most of the brown kelps of the ocean don't do too well in brackish waters, sea lettuce thrives down to an average salinity (my estimate) of as low as 15 parts per thousand and it will do nicely even if at low tide it is exposed to the air for an hour or two.

In its folds you will find a whole zoo of tiny fish, copepods, amphipods, isopods, decapods, gastropods and worms. At low tide the willets work over these green mud flats and at higher tides the ducks and larger fish have a go at them.

There are also some slimy grass-like water plants called *Enteromorpha* and some silky filamentous ones called *Cladophora* which again comprise many similar species. These algae are found along our entire coast wherever there is some protection from violent wave action. Seahorses twist their tails around these plants and eat passing copepods and baby shrimp.

OTHER-COLORED SEAWEEDS

Green flat seaweeds are found growing in brackish tidemarshes but the others, brown, purple, red and often lumpy, bumpy, or furnished with air bladders, are not. When they show up it is as a result of wave action and winds. These rock and tide pool and deep sea forms usually die and decompose after a few weeks in brackish mud-bottom tidemarsh waters.

PANIC GRASS

The weedy grass you know from roadsides and other dusty places is also found at the upper edge of the tidemarsh anywhere from Maine to Florida, growing near the marsh elder. This is a true grass, commonly called switchgrass or panicgrass and scientifically it is probably *Panicum virgatum* or *P. longifolium* if it is near tidewaters. It grows to a height of six feet but is generally only waist high. The flower cluster is anywhere from six to twenty inches long. The seeds are eaten by many upland birds and rails, teal and geese as well. Don't get too precise as you examine the illustration. It is precisely *P. virgatum*, but there are 149 additional species of *Panicum* east of the Mississippi and many of them might occur on your marsh. This genus of grass is perhaps the greatest single source of seeds for ground-feeding songbirds in the U.S. Just to give you an idea of how involved one can get in plant identification, be aware that there is a book entitled "The North American Species of *Panicum*", by Hitchock and Chase, Smithsonian Institution, U.S. National Museum, Washington, Published as Volume 15 in 1910. This monograph for the one genus runs 390 pages!

CATTAILS

The two most common tidemarsh cattails of our entire Atlantic Coast superficially resemble each other. Both grow to a height of eight feet and both develop flower clusters with the male part on top and the fuller female below. In more saline waters you will find the narrow-leaved cattail *Typha angustifolia*. Its two flower parts are separated by as much as several inches. These flowers are less than an inch in diameter.

The common fresher water version is called the broadleaf cattail or flag. Unfortunately there is also an iris which is sometimes called a flag, but "cattail" is a better name for *Typha*. This latter species is *Typha latifolia* and the two parts of its flower touch (or nearly touch) each other. Also, this latter plant is stouter. The leaves are wider and the flowers are as much as an inch in diameter. It is possible to find a line or a narrow band on some marshes where one species thins out and the other takes over. Of course this is true

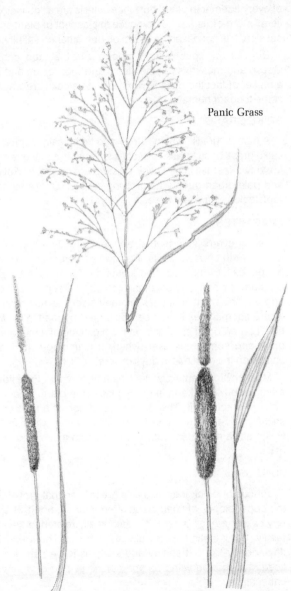

Panic Grass

Narrow-Leafed Cattail

Broadleaf Cattail

of every tidemarsh plant, but since cattails are a "collector's item" more of us are apt to notice this aspect of plant succession. It is not based on elevation but rather on salinity.

Both cattail species are important in the food chain. Geese and muskrats eat the stems and rootstocks; and as a matter of fact the male flowers and rootstocks are wholesome food for humans.

SPIKE GRASS

A marsh "grass" which is truly and technically a grass is spike grass or saltgrass, *Distichlis spicata*. It is hairless and up to two feet tall. The flower clusters are dense, spikelike and pale green during the growing season. Ducks eat the rootstocks, stems and seeds.

FRAGMITES

Ten or even fifteen feet tall and feathery at the top—this is the common reed grass, *Phragmites communis*. Some people call it fragmites. The stems are round and jointed like bamboo and the horizontal rootstocks creep on and on and on. This plant flourishes in many places as an "exotic" where tidemarshes have been disturbed by man. You will find it on dredged spoil and on the high side of a mosquito ditch and sometimes even where a clamdigger had his shanty on the marshes in days of yore.

Most environmentalists use it as a benchmark for a tidemarsh about to fail and in many places there are schemes afoot to eradicate it. This plant is not much of a food or a shelter to tidemarsh animal life but it competes with cattails which are a most desirable source of both these elements of life.

JUNCUS

Juncus or blackgrass is a little green rush that grows on the upper slopes of tidemarshes from Florida, north. Properly called *Juncus gerardi* it is technically related to the lily family, but it looks superficially like a grass. The flower is tiny and lily-like but so tiny and scaly as to be hardly noticed. It is mowed with *Spartina patens* as part of what we call "salt hay."

Spikegrass

Blackgrass

Fragmites

NEEDLERUSH

This is *Juncus roemerianus* and compared to black-grass *J. gerardi*, it has larger flowers on a sometimes larger plant. *J. roemerianus* is found from Maryland, south. The stems and leaves terminate in extremely hard sharp points.

THE SPARTINAS

The grass which has so much to do with the "meadows" of the tidemarsh is saltmeadow grass *Spartina patens*. Of the saltmarsh grasses this is the smallest and most deli-cately shaped species. Its small flowers and fruits are at-tached along one side of the stalks; rather than all around as in the case of timothy hay or cattails, for example. This plant is the source of most of the salt hay of commerce and its seeds will not germinate until they have been wet by salty water. All manner of animals eat the seeds, leaves, stems and rootstocks of *S. patens* as well as the other *Spartinas*. The roots of this plant and its larger counterpart *S. alterniflora* are what literally and figuratively holds the marsh together. The most wet of the tidemarsh grasses is the smooth cordgrass, saltwater grass, saltmarsh grass *Spartina alterniflora*. You will find it just about everywhere from Maine to Florida on the lower levels of tidemarshes. It grows to a height of nine feet and is extremely variable in size and form, but it is always coarser and larger than *S. patens*. The flowers, here too, are borne on only one side of the stalk. The stems are generally at least a foot under water at high tide.

The largest and most coarse of the *Spartinas* is the big cordgrass. Find it anywhere in brackish waters from Mas-sachusetts to Florida and call it *S. cynosuroides*. Notice that the flower clusters, sometimes over six feet above the ground, have thirty or more branches which are larger at the bottom than they are at the top.

Spartina pectinata is a big coarse tidemarsh plant which resembles *S. cynosuroides* except that the flower clusters have fewer than 20 branches and these branches are of equal size from bottom to top. This plant commonly called prairie cordgrass is found in tidemarshes between North Carolina and Massachusetts.

Big Cordgrass, S. *cynosuroides*   Prairie Cordgrass, S. *pectinata*

Saltmarsh Grass, S. *alterniflora*   Saltmeadow Grass, S. *patens*

Tidemarsh Plankton

Dinoflagellate

Radiolarian

Diatom

Planarian

Dinoflagellate

Dinoflagellate

Radiolarian

Dinoflagellate

Foraminifera

Copepod

Radiolarian

Radiolarian

Planarian

Diatoms

NOTE THAT THESE ORGANISMS ARE ALL MAGNIFIED
BUT NOT TO THE SAME SCALE

SMALL LIVING FORMS

Tidal marsh waters produce tremendous numbers of microscopic plants, and still other living forms which are intermediate between animals and plants; let's consider them as a group for convenience sake. Additionally, a few definitions are forced upon us here. These words crop up throughout the scientific literature. They are a part of the estuarine biologist's and oceanographer's jargon.

Green Plants consume chemicals and utilize sunlight and carbon dioxide to create cells of plant tissue. Most are green. Algae is just about the simplest plant, a single cell of algae can do quite nicely all by itself.

Dinoflagellata are protozoans which exhibit characteristics of both plants and animals. The name simply means that a single celled organism has two *(dino)* threadlike appendages *(flagella)*. Some contribute to the night-time luminescence of the water. Others, including the "red tide" are poisonous to man and are drawn into mussels in great numbers during summer time and if a large number are present and a person eats a large number of these mussels, that person may become ill, perhaps fatally so. That particular poison is *not* destroyed by cooking. Perhaps this is the origin of the old wives' tale about oysters being in season only during months which in English bear the letter "r." Some species of dinoflagellates, and there are many, have the ability to photosynthesize materials needed for their life process from sunlight, CO_2, water and chemicals dissolved in the water. Others capture prey and eat it. Others eat detritus. Others produce cellulose. And still others—one genus—can survive, and even thrive in several of these modes.

Animals don't utilize sunlight or consume chemicals but rather they consume plants or other animals or their wastes to create animal tissues. Very few are green, and none manufacture green chlorophyll.

Bacteria manage without sunlight, some even manage without oxygen; they are single celled. The cells never die unless injured but just keep dividing when conditions are right. Some cause disease, some consume wastes, some aid digestion and others help us to make cheese and sour cream. Although present in vast quantity in tidal marshes, they are individually apparent only to each other and to microscopists. The nitrogen cycle from animals to wastes to plants and back to animals depends on bacteria. When you see pale blue or gray scum in a stagnant mosquito ditch you are looking at bacteria. Some need oxygen to survive (aerobic) and some do very well without it (anaerobic).

Diatoms are plants which create match-box shaped two-piece silica shells to contain their single cells. There are in the sea, tremendous numbers of them. Their microscopic shells have created geologic strata hundreds of feet thick in many parts of the world. Perhaps they are the source of much of the petroleum oil in the earth today. There are two good reasons for mentioning diatoms in this book. First, they are a building block in virtually every food chain; and second they need silica to exist. Now, silica is the oxide of silicon (SiO_2) and it exists in nature as quartz and quartz-bearing rocks like granite. Silica is not especially soluble in anything except hydrofluoric acid or molten alkali, but somehow rivers do accumulate dissolved silica and they do carry it to the river mouths, estuaries and tidemarshes where the other minerals, carbon dioxide and nutrients necessary for diatom growth are available. So then, river mouths, estuaries and tidal marshes become a nursery for diatoms. Diatoms in turn are eaten by the larvae of many marine animals and many mollusks. Not all diatoms originate in tidemarsh waters but these waters do support tremendous numbers of them.

Foraminifera, like diatoms, are also single-celled. They are housed in globular or snailshell shaped one-piece shells which make repeated cell division difficult but not impossible. Surely not impossible since the famous White Cliffs of Dover are practically solid "foram" shells, deposited over the course of a few eons after which time the land there rose to expose the accumulation. Forams are not green; call them animals. They eat diatoms.

Diatoms are housed in silica (SiO_2) and produce diatomaceous earth as their shells accumulate, whereas forams create calcareous shells, chemically much like a clam shell

($CaCO_3$) and they make chalk.

Plankton is a catchall word. Plankton consists of "wanderers in the water," non swimmers and poor swimmers; diatoms, forams, protozoan animals like dinoflagellates, and all the small aquatic animals that eat the diatoms and protozoa. When we refer to *planktonic animals* or *zooplankton* we mean to say the animals that eat diatoms and similar plants. *Phytoplankton* is that part of plankton which is generally green and plantlike.

Detritus is slime or scum composed of decaying plants and the bacteria which contribute to this decay. Fiddler crabs and mullet seem to eat detritus. Debris is large particles of detritus. Large shrimp eat debris, smaller shrimp eat detritus.

The Rule of Ten. Examination of food chains and the efficiency of creatures who benefit themselves by eating others (or eating plants) brings to light an interesting bit of arithmetic. It turns out that for most organisms it takes ten unit weights of food to make one unit weight of themselves. The rule works out for all the steps in every food chain. A growing human being will gain a tenth of a pound (1.6 ounces) of tissue (bone, muscle, fat) by eating a pound of bluefish. The bluefish gained his pound by eating ten pounds of mackerel which gained those ten pounds by eating 100 pounds of anchovies which ate 1000 pounds of plankton which ate 10,000 pounds of diatoms and other small things. Incidentally, the other nine units become calories for expenditure of effort and generation of body heat rather than for creation of animal tissue.

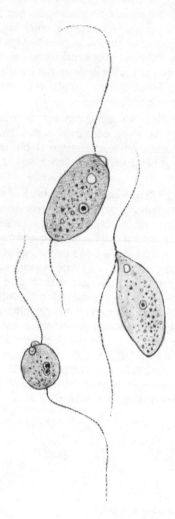

Dinoflagellates

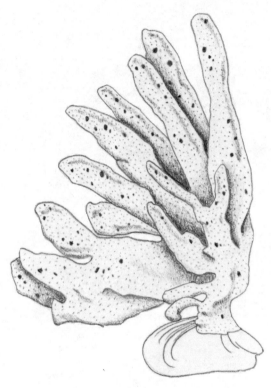

Sponge

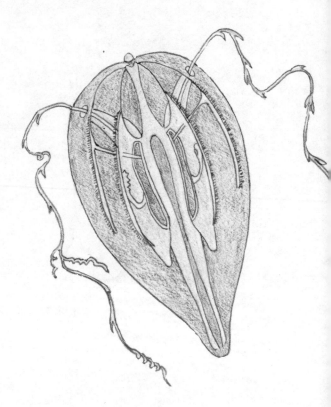

Ctenophore (*Lobata*)

LARGER LOWER ANIMALS

Sponges. Scientifically the phylum of sponges are called *Porifera,* the pore bearers. When we look at a sponge we are actually viewing a large colony of independent little living elements. A live sponge draws water and its food into its body cavity through tiny pores and it discharges the water and wastes through larger openings. The water is driven by flagella similar to those of the single-celled dinoflagellates. Some sponges take and retain their form from a skeleton made of microscopic "spines" of silica or calcium carbonate. Other sponges lack these minerals and are built up from a fiber or horn-like substance called spongin. The spongin sponge is the article of commerce and it is not found in tidal marshes. In fact, not many species of sponges adapt to the tidal marshes since the ever-present mud tends to clog them. These little living animal elements interlock or interface or fuse to form the mass we see as a sponge.

Coelenterata. Here is a phylum in the animal kingdom with three classes, all aquatic and most are denizens of salty water (estuarine or marine). Many coelenterates are luminescent and all have stinging nettles. They are carnivorous and are probably incapable of digesting plant matter.

First class in the phylum are the *Hydroids* which were made famous by that old high school biology laboratory favorite, the hydra. Other hydroids look like seaweed and still others in a sub-group called siphonophores are free swimmers like the Portugese man-of-war.

The second class are *Scyphozoa.* These are the jellyfishes, and most people recognize them as such. Sometimes a tidal marsh creek will be crowded with jellyfishes in late summer. The nettle-like stings of jellyfish can be relieved by applying Adolph's meat tenderizer, a papaya extract which seems to digest the protein poison. Be sure you get the tenderizer which is not salted or spiced.

One jellyfish, *Chrysaora quinquecirrha* is so common and so nettlesome (it is called the stinging nettle) in Chesapeake Bay that it is considered to be "probably the single most important factor limiting recreational use of the bay." My reference here is Chesapeake Science, Volume 13. December 1972.

The third class in the phylum of *Coelenterata* is *Anthozoa* and this includes the anemones, sea pens, sea pansies and corals. Anemones and sea pens are sometimes found in tidemarsh mud. Corals are rare north of Latitude 30° N, and are not part of the population of a tidal marsh, anyway.

Ctenophora. Ctenophores resemble jelly fish. They have no skeletons, no stinging nettles, no appetite for plant life. They are carnivorous and luminescent; they appear in vast schools and they certainly resemble jellyfish. Some are reported to eat oyster larvae. They range in size from a melon down to a pea. *Ctenophores* are phylum all by themselves today but older texts consider them as a subphylum of *Coelentrates.* Careful examination of *Ctenophores* shows them to have more well developed nervous systems than *Coelentrates* but for most tidemarsh visitors the only outstanding difference is that while a *Ctenophore* resembles a jellyfish it does not sting like one. Other names include: Water gulls, comb jellies, sea walnuts, jellyfish. Their bodies are 95% water and less than 1% animal tissue (which is mostly protein). The remainder is mostly salts.

Platyhelminthes. These are the flatworms. One group, the *Turbellarians,* are found in tidal marsh waters and the others *(Trematodes* and *Cestodes)* are mostly the parasites of other animals. The *Trematodes* include the liver flukes, and the *Cestodes* include the tapeworms.

There is but one opening to the body cavity The class *Turbellaria* includes some free-living flatworms which have, inside their bodies, colonies of algae. The plants reproduce and guarantee food for the worms by photosynthesis, and the worms (called *Acobea*) digest some but not all the plants. Another order within the *Turbellarians* is called *Rhabdocoles.* These worms may be only 1/8" long and are very common in tide marsh channel mud.

Another order, the *Tricladida* includes the *planaria.* This creature has a real alimentary canal with three branches. The *planaria* is famous because it can be taught to perform a "trick" and then if it is cut up and fed to another planarian this second animal will automatically perform the same trick. *Planaria* are visible to the naked eye and are often seen creeping over and through marsh detritus.

A fourth order of *Turbellaria* are the *Polycladida.* These worms have branched digestive tracts and some grow to a length of six inches. In your travels you probably notice a relatively common Turbellarian, the white flatworm *(Bdelloura),* which is associated with horseshoe crabs. This worm is found creeping over the crab's shell where it picks up scraps of food.

ANNELIDA

The phylum Annelida (also called *Annulata*) takes in the worms whose bodies are segmented. That is, they are structured of rings and frequently each ring is equipped with unjointed appendages which help them to creep or swim. There is no shell, no hard parts except perhaps some needle-sharp teeth described in scholarly texts as "horny jaws." Some species, when cut up, are capable of regenerating perfect worms from each section, and some others can regenerate only their proboscis or their rear portions.

CLAM WORMS

The best known tidemarsh dwellers in the Annelid phylum (and there are many, many species) are the sand-worms, clam-worms, blood-worms, and lugworms used as bait by sport fishermen. Be forewarned that in various parts of this country the names sand-worm, blood-worm, and clam-worm refer to the same or very similar species.

The clam-worms, genus *Nereis*, you will probably encounter are: North of New Jersey - *Nereis virens*; north of Virginia - *Nereis pelagica*; New England to South Carolina - *Nereis limbata.* All are active at night when they swim about like eels, attacking other worms, crustacea and probably anything else they can subdue. This is also the time that eels and striped bass and white perch eat the worms. During the day *Nereis* is found in burrows in the mud; and some tidemarsh mud flats are intensively worked over for these worms for the bait business. A seven or eight inch *Nereis* usually retails for two or three times the price of a large earthworm (night crawler).

Extended Proboscis of Bloodworm

Nereis are usually a dull-green or olive-green. Sometimes there is irridescence. The head is furnished with two horny jaws or hooks and I was always under the impression that they not only bite but also sting. That is, they might inject some poison into the penetration achieved by the bite. As a boy, I can remember that my fingers would swell up and be sore whenever I was bitten by a worm, but I find nothing in the technical literature about worm venom.

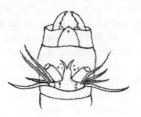

Proboscis of *Nereis*

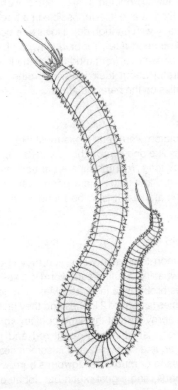

Nereis, the Sandworm

The other large common mudflat worm is popularly called the blood-worm. It is one of several species of *Glycerids* and it is much like the *Nereids* except that there are not two but four hooks making up the proboscis. Good fish bait, but really nasty. The color is more pink and rose and the body tends to be rounder with shorter parapodia. You might prefer to call those parapodia feet or bristles. When you cut up a blood-worm, it bleeds a lot of red blood. Like the *Nereids* it burrows and is also found swimming at night. This swimming may have something to do with its reproductive cycle.

There are hundreds of species of mud-dwelling predatory, horny jawed, nocturnal segmented worms between Maine and Florida and it is impossible in a book of this size to describe any with enough detail to assure positive identification. If you need a few for bait, they can be dug with a potato fork or a shovel from mud flats at the lowest of tides. Just be careful about their jaws. Still other worms have stinging nettles on the parapodia.

CAPITELLA

The tidemarsh worm that looks most like an earthworm and behaves like one is probably *Capitella capitata*. It is up to five inches long, red and tapered at both ends. You will find it near the low water mark under waterlogged wood and similar materials. It is to be found on the entire U.S. East Coast.

LUGWORMS

The common tidemarsh worm you may encounter that looks somewhat like a tufted earthworm is a species of *Arenicola* and is probably a lugworm. Maximum length of the longest species is about 12 inches and they lack the leglike bristles and horny jaws of the sand or clam worms and the blood worms. Also, the irridescent red and green clam worm is larger, is a swimmer, and is sometimes found in a thin tube in sand or mud. The lugworms burrow constantly, like earthworms, and ingest much but not all of what they tunnel through. The organic matter is digested and the inorganic sand and mud which accidentally enters is simply passed through its alimentary canal. Whereas the clam

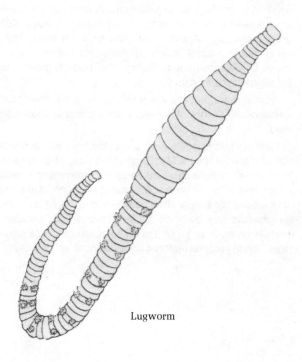

Lugworm

worm carefully selects individual items of its food and captures them, the typical *Arenicola,* the lugworm, is a less discriminating feeder, swallowing much of what he tunnels through. If you dig in the tidemarshes, you will eventually find some, and they do make good fish bait, but many casual visitors never see one. Illustrated is a common lugworm *Arenicola cristata:* it is colored a muddy-purple-blue-black and marked with small light-colored fibrous lumps which are actually gills. Another common lugworm found from Long Island Sound, north, is *Arenicola marina,* and it grows to a length of eight inches. The body is green and the tufted gills are brownish red.

Bear in mind that this *Arenicola* family is but one of the tremendous phylum of annelid worms found in tidemarsh mud.

I feel obliged to quote from one of the best natural histories of marine animals, MacGinitie: "We do not like to leave the annelids with so little said about their interesting habits. Unfortunately, what we know about them is not complete life histories but simply little notes here and there that make an unsatisfactory and incomplete picture of their activities. There is no more virgin field of investigation open to a naturalist than the natural history of marine annelids."

MOLLUSKS, ALSO SPELLED MOLLUSCS

There are 121 recorded species of mollusks in Chesapeake Bay alone. There are 154 references to them cited by H.T. Pfitzenmeyer in Contribution No. 509, Natural Resources Institute, University of Maryland, Solomons, Maryland 20688. Some of these references are in themselves bibliographies! You see, there is no shortage of information about the natural history of the mollusks; the big problem for most casual readers is that of how to glean the particulars that interest them without sinking into the morass of data. Once you go into those references you are hooked. You are then no longer a "casual visitor." Condolences.

This phylum has five classes, all of which have one thing in common. They have soft bodies. In fact, *mollusca* means soft. Don't be misled by this nomenclature business; a raw conch or a raw abalone is about as soft as a piece of automobile tire tread, with all four plies left in. In the clams and chitons and most snails, there is a hard external shell, but in most cases, the body has no internal structural stiffeners. There is no jointed leg as in arthropods, nor is there a spine or spinal chord as in the chordates. Mollusks do however have many of the organs found in man. Some have highly developed sense organs, hearts, stomachs, rectums, eyes, and real thinking brains, and some don't. Some mollusks are vegetarians, others are predators and still others are scavengers or parasites. From where we stand, the octopus seems to be at the top of the heap.

Of the five classes those most prominent on tidemarshes are:

Gastropoda (the stomach is the foot)—these are snails and whelks and periwinkles. Generally there is but a single spiraled shell (univalve). They have heads and eyes and tentacles. Some are provided with a horny front door called an operculum and some are not.

Pelecypoda (the foot is shaped like a hatchet)—these are the double shelled or bivalve mollusks including clams, oysters, scallops, and mussels, all found on tidemarshes or in tidemarsh waters. So, by now, you know that the word used by technical zoologists for shells is valves. Bivalve mollusks are also known as "filter feeders."

Cephalopoda (the feet are attached to the head)—these include the squids and there is but one truly brackish tidewater species which you should regularly seine from marsh creeks and another which may stray in from time to time. Some cephalopods do have internal stiffeners. The squids, for instance, have a long transparent horny "bone" within the mantle, but this is considered to be an internal shell since as we know, mollusks are boneless.

SNAILS

These are the gastropod mollusks whose name is derived from the fact that the stomach is in the foot. Mollusk = soft, gastro = stomach, pod = foot. Limpets are also gastropods, but their spiral-shaped larval shells are modified once they settle down. Getting back to the snails of the tidemarsh, there are only a score of species, but their abundance in some places is staggering. On tidemarsh flats wet perhaps half of the time, you may discover that the *Nassarius* snails are so thick your foot prints will cover a dozen at every step. Some but not all snails are equipped with a plate, called an operculum, sized precisely to close off the only opening to the shell.

Snails obtain their food by the rasping action of an organ called radula. Some species use the radula to scrape algae off rocks and others use it to drill through shells of other mollusks in order to eat them. Many species of snails are hermaphrodites. That is to say, each snail is equipped with a complete set of male and female organs and so any two can function sexually as four, but any one cannot fertilize itself.

The tidemarsh snails with thick basketweave shells, opercula and fairly long spirals that frequent the mud around the stems of *Spartina* and *Juncus* are probably a *Nassarius* also known as *Ilyanassa* species; and the common mud snail is the most likely candidate. This one inch or smaller mud flat dweller is *Nassarius obsoletus,* sometimes also referred to as *Nassa obsoleta.* It eats algae and is an opportunistic scavenger as well. Mud snail shells are not easily crushed and so this animal is not frequently preyed upon by birds or fish.

If the tidemarsh snail you see is relatively thick-shelled with a short spiral and an operculum, it is probably one of six or so species of periwinkle, *Littorina*. Periwinkles are esteemed by Europeans as food and are prepared by boiling, roasting or broiling. They have a piquant nutty flavor. A large periwinkle is a bit smaller than a bing cherry. Periwinkles are more commonly found in rocky places, but tidemarshes are not without them.

Another smaller, thin-shelled operculum-bearing genus found in marsh creeks on sea lettuce (*Ulva*) and plant stems like *Juncus* and *Spartina* is probably a *Hydrobia* species and a common typical form of it is *Hydrobia minuta,* the swamp hydrobia found in tidemarsh pools in New England and as far south as New Jersey.

The other candidate for small thin-shelled snails, this one without an operculum looks like the common freshwater pond snail. It is properly called the salt-marsh snail and it is found along our entire East Coast. Scientifically it is *Melampus bidentatus* and a large specimen is but half an inch long. Many birds eat them with avidity. When the tide rises, this species frequently climbs plant stems to keep from being totally immersed. The inner lip of the shell opening has two indentations and these give it the name *bi* = two, *dentatus* = teeth.

Salt-marsh Snail, *Melampus*

Swamp Hydrobia (also known as
Minute Seaweed Snail),
Hydrobia minuta

Periwinkle, *Littorina*

Mud Snail, *Nassarius*

SCALLOPS

The wider tidemarsh creeks, slow moving and shallow bays are often inhabited with scallops, the mobile bivalve with the single large delicious adductor muscle. All the soft parts are edible, raw or sauteed or broiled or fried, but usually only that famous creamy white muscle is served to diners.

When you see an empty shell its identity poses no problem. The family is called *Pectinidae,* and the common bay scallop (*Pecten irradians*) is found from Cape Hatteras to Cape Cod. It grows to three inches in diameter and the 17 to 20 interlocking radiating ribs, the conspicous wing and the orb shape all make it easy to recognize.

When you see a live scallop moving in its jetstream a yard or so with every squirt, with its multi-stalked blue eyes and its rapid motion, you may not believe what you are seeing. Especially if you assume that when it claps its valves shut it will always shoot off in the direction its hinge and wings are pointing. Not so—it pumps its water jet out between its wings and it advances at high speed toward its blue-eyed opening. This information is good for barroom wagers. The same scallop is also capable of moving toward its hinge, but not as rapidly. This information may engender bar-room fights over wagers.

From Cape Cod to the north this species peters out, probably because the tidemarshes, creeks and bays freeze so hard and deep; but to the south there is another shallow-water form, *Pecten dislocatus,* the southern common scallop and it is found from North Carolina to Florida. It is smaller than its northern relative *P. irradians,* being at most only two inches in diameter and there are many more ribs, closely spaced.

MUSSELS

Our Atlantic Coast supports only a few species of mussels. The best example of a genuine tidemarsh mud dweller is that solitary mud mussel, the ribbed mussel, and it is found alone in a hole in a creek bank or on a low flat. Its scientific name is *Modiolus demissus plicatulus* and it may

Scallop

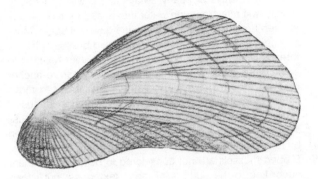

Ribbed Mussel, *Modiolus*

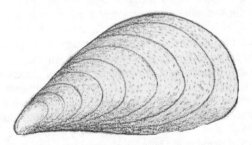

Blue Mussel (also known as Common Edible Mussel),
Mytilus

be under water at all times or may be submerged only at the higher end of the tidal range, but since it eats and reproduces only when submerged, it generally does better when it is always or nearly always under water. A ribbed mussel will retain enough water if exposed by the low tide to keep breathing until the rising tide covers it again. Mussels can, at any time, cast out hairs called byssus or byssal threads which anchor them to the peat in their holes or to some other substrate. With the byssus they can re-anchor or even move a little, very slowly. This ability to move seems to be true for all mussel species. The ribbed mussel is found from Georgia north to Prince Edward Island. Mussels eat diatoms, algae and dinoflagellates. They are eaten by rats, muskrats, crabs, borers, starfish and some ducks. Ribbed mussels are not considered to be edible by man, according to the authors of many textbooks, but I feed them to my family occasionally. We thrive and enjoy them.

Probably the reason the solitary ribbed mussel is not favored is that there are so many *Mytilus edulis* so readily available in convenient clusters. This latter species is actually called "the common edible mussel," and by its very name it tends to exclude the others. Very unfair. *M. edulis* is found in the saltier waters, pure sea water seems to be its usual habitat, but the ribbed mussel is much more tolerant and is found in many areas of lower salinity. When I was younger and more scientific I tried to put numerical values on things like tolerance levels, but after long reflection I realize that reliance on very precise measurement in these fields of natural history is not especially rewarding. Look at yourself. Exactly how much noise can you stand? How much whiskey can you handle? How much radiation can you tolerate? It is easy to measure and record noise and whiskey and radiation. It is not at all easy to define "stand," "handle" and "tolerate." It is very easy to define a tidemarsh in terms of the species of plants we find growing on it. It is virtually impossible to define the boundaries of a tidemarsh by arbitrary tide range or salinity numbers. Look at the Amazon River—potable water 300 miles off shore. Look at the Hudson River, tidal but not saline to Albany, 150 miles north of the Statue of Liberty. When you visit a tidemarsh you may find a creek with a large population of ribbed mussels halfway between its mouth and its fresh water source. The density of population will peter out as you move in either direction. Downstream there might be some clusters of seagoing *M. edulis* or even horse mussels *Modiola modiolus*, upstream there might be some fresh water mussels, muscling in.

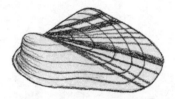

Tulip Mussel, *Modiola*

The small southern form, found from Delaware Bay to Florida is *Modiolaria lateralis* and it is not especially "mussel shaped", but it does spin a fibrous nest of byssus and this pretty much makes it a mussel. The maximum size is one inch. Another southern form is *Mytilus exustus*, the scorched mussel. It grows to an inch and a half and is frequently found attached to another mollusk. It is common from North Carolina to the Gulf of Mexico.

The two inch southern form of the ribbed mussel is *Modiola tulipa*, the tulip mussel. It is prone to attach itself to a piece of submerged wood and since it is distinctively shaped and marked, it is not easily mistaken. The shells are yellow-brown with darker brown stripes.

OYSTERS

This bivalve molluscan pelecypod filter feeder has many names, probably because it has been closely associated with man for such a long time. The family is *Ostraeidae* and the genus *Ostrea* (or *Ostraea*) was officially described by the number-one describer and classifier and namer, Carl von Linne (1707-1778) who we generally know by his latinized name *Linnaeus*. The Greek word is ostreon (from ostrakon) meaning "hard shell." In the Orient, oysters were being farmed while those Greeks were still running around in fig leaves.

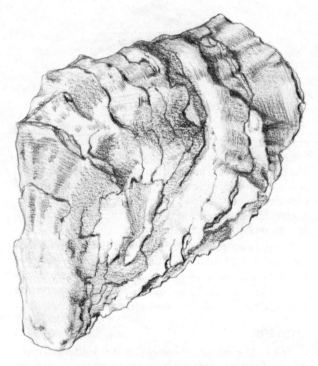

Virginia Oyster

The coast of the U.S. supports but two native species, and no one should have trouble identifying an adult oyster. Both of its shells are rough and irregular and dissimilar. The left shell is convex and larger than the right shell. Most oysters are found attached to something, a rock, another oyster, a shell, or a mangrove root; and it is the left shell which makes the connection. Oh yes, when you hold a bi-valve in your hands, the right shell (or valve) is in your right hand when the hinge is on top and the beak or bill points toward you. Admittedly it may be difficult for you to find the beak on an irregularly shaped oyster. There is but one ad-ductor muscle in an oyster whereas a typical tidemarsh clam or mussel has two of these powerful clamping springs of muscle tissue. The Virginia oyster, *Crassostrea virginica* is our East Coast species and is found in bays, inlets, and estuaries from Maine to Florida and throughout the Gulf Coast. It cannot survive long in fresh water. It needs a mini-mum of ten parts per thousand of salinity and some geo-graphic races thrive in pure seawater. It needs a surf-free bottom and it prefers to be attached, lest it sink into soft mud and suffocate or starve. It is most active at about 80°F. and it will feed only at temperatures above 40° F. Pro-longed freezing kills any oyster. Oysters spawn at 74° F. and so in some waters there are two spawning periods, one in early summer, and another in early fall. In other areas to the north there is but one spawning when the water is war-mest, and in the South, oysters spawn nearly continuously.

Oysters need tremendous quantities of lime to create their shells and young seem to choose shell to settle on for this reason.

This is terrible! One wonders how with all these limita-tions and restrictions they have managed to leave more than a fleeting, obscure fossil record. Well, for one thing, oysters are very sexy. They are mature in perhaps two or three years, depending on their growth rate and they func-tion mostly as males during their first reproductive year. Older larger specimens are mostly females, but they can switch from time to time. Hermaphrodites exist, but *they* usually function as females.

So, looking at the bright side from the standpoint of an oyster, it's versatile in its niche. It lacks mobility, but the tides deliver its food, disperse its eggs and remove wastes. The food supply may dry up with a low tide or be depleted in cold weather; the oyster closes his shells and lives off stored fat. And, oysters are prolific. Prolific is a word that understates the most conservative estimates. The figures from various authorities do not always agree; but for starters consider a female that functions as such for ten years. Each year she will release 60,000,000 to 300,000,000 eggs. If just sixty million eggs were fertilized and set and grown to maturity and harvested and shucked, it would take the output of only fifty such female oysters and perhaps the same number of males to provide us with our entire current U.S. consumption of about 3.0 to 3.6 billion (3,600,000,000) individual oysters. And, in addition, the same females might also spend some part of each of those ten years functioning as male and fertilizing some other female's eggs.

Oysters have no heads and no mouths; they eat and breathe with their gills. A large oyster will draw as much as three gallons of water per hour through its gills and extract diatoms and dinoflagellates for food, and oxygen for its metabolic processes like tissue building, work and reproduction. Technically, the opening to the alimentary canal is termed the mouth, and this they do have; but they lack the sort of mouth we find on a fish or even a small snail.

Some oysters are found in the intertidal region—above the low mark and below the high mark. This lifestyle is possible only where there is no extended period of intense cold, and oysters are found in these intertidal areas only between Cape Fear River, North Carolina, and northeastern Florida. In these parts, the tide ranges between four and eight feet and the oysters do quite well. They are out of water for periods up to five hours and are subject to cold air and even near-freezing temperatures, but on the other hand they are able to close up and wait while their enemies are not so adaptable. True, they miss some eating time, but they gain from reduced losses to starfish and drills.

I keep coming back to their life style because it is so frequently linked to the waters adjoining tidemarshes and because so many other tidemarsh life forms are tied in with the oyster. First let's taste the water. It will be brackish and about about 74° F. when the oysters release their eggs and sperm or fertilized and developing eggs. (Some authorities tell us that the female holds her eggs and draws in the sperm, and that union takes place within her shell and that she releases larvae and not unfertilized eggs.) Regardless, the union produces larvae which measure less than one five hundredth of an inch across. Under the microscope they look clam-like. Each microscopic creature has a foot like a clam and hairy (cilia) processes on the foot which help it to swim. It eats single-celled plants and animals and it grows to perhaps two millimeters (0.08 inches) across its shell (now the larvae is called "spat") before it settles on something solid (this substrate is called "cultch") and attaches itself permanently. The set of the spat on the cultch (oyster men's jargon) is sometimes triggered by an increase in the copper ions in the water. When conditions are right, it leads to crowding so fierce as to push some oysters off the substrate as they grow.

The Virginia oyster is the only native member of its genus found along the entire U.S. East Coast. However, from North Carolina to the Florida Keys there is another oyster species; *O. frons,* commonly called either the tree oyster or the coon oyster. This species is prone to cluster in the roots of mangrove trees. They are a favored food of raccoons. *O. frons* are never as large or desirable as *O. virginica,* but they are good to eat.

Oysters are the epitome of human gastronomy. No other edible substance holds a candle to the oyster for universal appeal. Truffles and sturgeon caviar may cost much more per pound but they don't have the universal appeal. There have been poems, songs and recipes written for oysters in virtually every language. Oysters also occupy an important place in the economic system (from which we derive the word ecology) of tidemarshes since they eat tremendous quantities of the diatoms and dinoflagellates which originated in the tidemarsh waters. Then, in turn, oysters are food for oyster drills, starfish, horseshoe crabs, and many species of fish. The plankton eaters including alewives, menhaden, jellyfish, hydroids and shrimp consume the larvae. The strong jawed fish attack the juveniles and perhaps even the adult oysters. Blue crabs, starfish, raccoons,

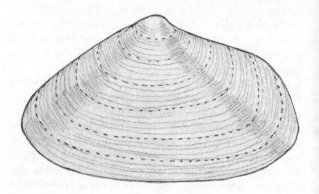

Soft Shell Clam, *Mya*

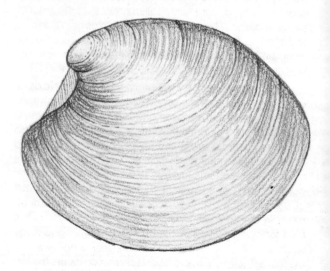

Hard Clam, *Mercenaria* (*Venus*)

muskrats, and mud crabs take their toll of adult oysters; and if we are to believe Lewis Carroll, so do the walrus and the carpenter.

Tidemarsh oysters are harvested from bay and creek bottoms. They are usually attached to shells and wood, but occasionally they are free. Oystermen rake, tong and dredge them from the bottom and they are frequently scraped from piling, docks and other man-made structures. The "R" month consideration is true for gourmets, but actually any oyster taken from unpolluted waters is edible all twelve months of the year. To open an oyster, a novice should knock or clip off enough of the thin bills of the shells to be able to insert a knife into the soft part. There the knife should be swept across the middle of the shell to cut the adductor muscle, whereupon the oyster springs open.

Oyster shape and color is a function of the bottom and the food. An oyster which is sinking in mud as it grows heavier will grow in length to keep its bill above the mud. This is self-defeating but what else can he do? Oysters in algae filled waters are greenish and still other colors have been explained as environmental rather than genetic in origin.

CLAMS

There are many species of clams in the sea but the mud-bottom-tide-marsh-creeks support large populations of relatively few species. Clams are able to move in mud or sand and some, like the razor clam move so rapidly as to sometimes defy capture. Others, like the cherrystone are more sedentary, but certainly capable of adjusting their position in creek bottom "clam mud" to keep from suffocating and to find filterable water from which to extract food.

SOFT SHELLED CLAMS

When you walk on stiff mud at low tide anywhere from North Carolina to the Arctic and see squirts of water shoot up before you, the soft-shelled steamer clam *Mya arenaria* is probably underfoot. Also known as the long clam, the soft-shelled clam, the long-necked clam and the sand clam. You may think of some more vulgar names for it, other people already have. It is usually buried in stiff mud so that only its long siphon sticks into the water and it is frequently exposed at low tide. It is good to eat by man and beast alike and is also used as fish bait. A siphon tip is sometimes broken or bitten off and if everything else is intact, a new siphon tip is soon regenerated. This species is steamed, and washed in its own broth to shake off the sand and then dipped in butter for high class eating. It is not eaten raw because it is sandy and tough.

North of Massachusetts a shorter, wider clam, appropriately called the short clam, *Mya truncata*, takes over. In its lifestyle it is much like its southern cousin.

HARD SHELLED CLAMS

The hard clam *Mercenaria mercenaria*, formerly known as *Venus mercenaria*, a prettier name, I think, is more desirable as food than is *Mya. Venus* has a smaller siphon and it is found in deeper water where people rake them or tread them out of soft mud, while *Mya* is dug with a potato or manure fork or a shovel. *V. mercenaria* is commonly called the round clam, the little neck, the cherrystone, the hard-shell, and unfortunately, the quahog. Quahog is an unfortunate name because in some places the surf clam is also (although incorrectly) called quahog.

V. (or M.) mercenaria is the wampum clam; it is the purple part of the shell that the Indians prized. This mercenary aspect is probably the reason Carl Linneaus named it as he did. Pearls are sometimes found in these quahogs but they are valueless. A larger southern form *M. campechiensis* is found from Texas north to Virginia.

Small specimens are called cherrystones and these are eaten raw. Larger specimens are used in chowders or are baked. Really large shells end up as ashtrays. The wampum trade is just about washed up.

RAZOR CLAMS

The shell is shaped and sized about the same as the straight razor now-a-days used mostly in barbershops. This family of clams, *Solenidae*, is properly called razorshell clams and there are two species that really look like razors and two additional which look less so but have similarly arranged soft parts. These clams are so adept at advancing through sand or muddy sand that diggers are hard put to catch up.

The green razor-shell is never more than two inches long. The hinge is straight and the shells are too small to accommodate the soft parts, so they stick out at one end or the other or both ends. This is *Solen viridis* and it is common in sand and muddy sand from Florida to Rhode Island.

The common razor-shell clam *Ensis directus* is also called the sword razor-shell and it is truly common for the full length of our Atlantic coast. Maximum shell size is seven inches, but the soft parts stick out of the ends. The shell is olive green colored and curved, whereas the green razor-shell is straight. This species is delicious fried and is a popular food when obtainable. No one I know has really mastered an easy technique for capturing these fast-moving mollusks.

Common Razor Shell Clam

The last two "technical" razors you may encounter look more like soft shelled steamer clams than real razors and they belong to the genus *Siliqua*. Both are smaller than three inches and are commonly called pod shells. *Siliqua costata* is the shallow water form and is found from North Carolina, north. The shells are smooth, glossy, and violet blending to olive and pale yellow-green. The second of these pod shells is mentioned in passing because sometimes you will find the shell in New England tidemarshes where it washed in during a storm. This is *S. squama* the scaly pod shell. It is thicker than *S. costata* and yellow-green lacking violet.

Fragile Razor Clam, *Siliqua costata*

FRESH-WATER CLAM

Sometimes called a wedge clam (which is confusing since there is another genus *Donax* with that common name), the fresh water clam *Rangia cuneata* is native to Alabama and southwest to Texas. This species was introduced to Chesapeake Bay where it is now a dominant mollusk species and it makes up a substantial part of the animal life (biomass). That is to say they are very abundant in many areas.

This clam deserves your notice when you find it from Virginia, south, not only because it is a well-established exotic but also because one of its common names suggests freshwater but in fact it also thrives in brackish waters. The female retains her fertilized eggs until they hatch into parasitic swimming shelled larvae called *glochidia* which attach to the fins or gills of fish. There they remain during their early development. This lifestyle tends to disperse their population. They appear as barely visible specks until they fall off and then they begin to act like what *we* expect clams to act like.

Common Rangia

SQUID

Among the mollusks there is but one cephalopod (its feet = pod are attached to its head = cephalus) which makes its home in tidemarsh brackish waters. This is the squid and if it is less than three inches long (mantle length) and has rounded fins and you found it south of Delaware Bay, it is probably *Loligo (Loliguncula) brevis*. A larger species, not so prone to enter brackish waters is *Loligo pealii* and it grows to perhaps a foot mantle length. Its fins are more nearly low triangles. *L. pealii* is *the* common squid from Hatteras to Massachusetts.

Both species travel rapidly in either direction by twisting their jet discharge tube. With their ten tentacle-tipped legs, large eyes, parrot-like jaws, flashing opalescent color and occasional inky discharge you cannot mistake them. There is some evidence that this inky discharge has a paralyzing effect on enemies or prey and several cephalopods are reputed to be poisonous and certainly any live one is capable of biting you. These creatures are formidable predators on small water life and in turn they are eaten and relished by perch, fluke, striped bass and bluefish.

Male squids inseminate females with a packet of sperm. Eggs perhaps two inches long and a half-inch in diameter are then laid in clusters. Each cluster—usually attached to a plant—might contain several score eggs packed in an inedible jelly.

One close relative of this ten-legged squid is the eight-legged octopus, but none of them inhabit our tidemarsh creeks.

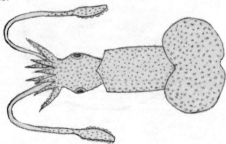

Squid, *Loligo brevis*

ARTHROPODS - CRUSTACEANS

In all animal life there is no more diversity, or numbers of species, than you can find in the *Phylum Arthropoda*. In fact, there are five times as many species of arthropods as there are in the entire remaining animal kingdom. They swim and fly and creep and crawl and burrow. Their specialization boggles the imagination. The prominent Classes in the Phylum are sometimes arranged, in older books, as follows:

Crustacea-	lobsters, shrimps, crabs, isopods, copepods, amphipods, water fleas, and barnacles.
Arachnoidea-	spiders, mites and (to confuse you a little), horseshoe crabs, which are not crabs at all.
Onychophora-	perpatus-velvet worms, only a silly millimeter long.
Myriapoda-	millipedes and centipedes. Some authorities make two classes (*Diplipoda* and *Chilopoda*) out of this.

Hexapoda or *Insecta*-insects.

A more modern arrangement is: *Crustacea*
Arachnida
Insecta
Diplopoda
Chilopoda
and several others

Representatives of all these classes are found on or bordering on tidal marshes and it would be appropriate here to mention that there are nearly a million *known* species of insects alone! Be humble. All adult arthropods are furnished with jointed legs and none have a dorsally located equivalent of the vertebrate spinal chord. Your chances of finding an arthropod unknown to science on a tidal marsh is good. Your chance of knowing positively what it is or even if it has already been described is poor. For one example, R.W. Miner in his *Field Book of Seashore Life* provides illustrations of about 50 species of Amphipods on our North Atlantic coast alone, and Amphipods are but a small order within the Class *Crustacea*.

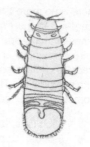

Gribble, *L. lignorum*

The common gribble of the Atlantic Coast tide-marshes is a relative of the equally common pillbugs, woodlice and sowbugs of our basements and gardens and woodlands. Although many people lump pillbugs and their ilk with the insects, this is not at all correct. All the aforementioned are in the:

Order Isopoda—Their legs are all similar, some are terrestrial, some are amphibious, some are aquatic.

Class Crustacea—Shelled arthropods with no wings, with two pairs of antenna, with no sharp separations between head, thorax and abdomen.

Phylum Arthropoda—With jointed appendages.

Kingdom Animalia—Capable of movement, incapable of photosynthesis, they eat plants and other animals and their wastes.

Insects, which are yet another class in the great Phylum Arthropoda have three pairs of legs, two pairs of wings, one pair of antenna, and their bodies are typically in three segments—head, thorax and abdomen. The legs are attached to the thorax. There are exceptions. This is to be expected, considering that we know that there are about a million species.

ISOPODS

Visitors to a tidemarsh sometimes confuse the order of amphipods with another crustacean order, the isopods. The pictures should help but, briefly, the amphipods are relatively flat from side to side and the isopods are flat from top to bottom. Our tidemarsh isopods are all less than an inch long, all are scavengers, all are found in marsh grass, and marsh debris. The females carry their eggs in brood pouches between several pairs of legs. Isopods are so named because *iso* = same, *pod* = foot; their feet, usually seven pairs, are all approximately the same size, and functionally similar.

Tidemarsh isopods look like wood lice, sow bugs and pill bugs. Some species are terrestrial, some semi-terrestrial, and others wholly aquatic. Swimming is accomplished underwater with their feet down. There are an awful lot of them around and all sorts of larger creatures eat them avidly. Incidentally, if they get into your clothes, they may nibble at your softer parts. One typical isopod is *Philoscia vittata*, a third of an inch long, and mottled brown with yellow stripes. These tidewater forms are amphibious and they must stay damp to breathe but they strive to avoid being submerged for long periods at high tide. Perhaps this is not a breathing problem, but rather an attempt to keep from being eaten by fish which hunt through the stems of submerged grass at high tide. The trouble for isopods is that when they climb out of the water at high tide to avoid fish bite, they become vulnerable to bird bite.

Another isopod of the tidemarsh is *Porcellio rathkei*. It grows to a half inch in length and favors higher land than does *Philoscia*. It is more terrestrial in habitat but still needs moisture to keep its external gills operative.

Perhaps the best known tidemarsh isopod is *Limnoria lignorum*. It has what I believe is a really wonderful common name, the gribble. The gribble is ubiquitous on tidemarshes. It is found on our entire Atlantic coast, from a little above the high water mark to a little below the low tide water mark. This tiny creature (it is always less than a quarter of an inch long, and most are less than half that) eats wood, dead wood, boat bottoms, piles, wharves, docks, dikes,

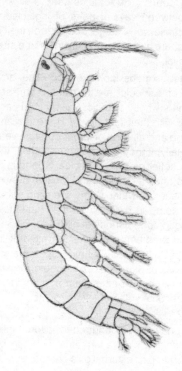

Amphipod, *Gammarus*

and of course flotsam. Gribbles can consume floating or sunken driftwood at the rate of an inch of depth per year. The trouble is that although they remove obstructions to navigation from our waterways, they do not discriminate, but go after valuable construction with equal avidity. The Port of New York was for many years virtually free from marine borers like the gribbles and the teredo, but with the recent reduction of pollution these little fellows are again establishing themselves.

Other isopods scavenge and still others nibble people who stand around in isopod country. Some isopods are commensal—they "eat at the same table"—with other creatures, and still others are parasitic on fishes and are found in their gills or mouths. Some species of isopods are parasites on other species of isopods. It would be easy for a person to spend a lifetime studying isopods, and if he were short lived or a slow learner he would not even catch up with what is already known.

AMPHIPODS

Amphipods are crustaceans that look like little shrimp. They are semi-terrestrial, and when they are not in the water, they are generally in damp shaded places. This is necessary for their survival since their respiration depends on gills which are affixed to the inside surfaces of their legs. Amphipods are scavengers. They eat dead plant and animal matter on tidal marshes, and are also useful to naturalists who need complete study skeletons of small delicate specimens. A small dead fish or perhaps a mouse is simply fastened to a board or wrapped loosely in a coarse mesh or placed in a jar with a mesh opening and secured near the waterline in a tidemarsh. The amphipods and isopods will make short shrift of the soft parts and the scientist has merely to collect the bony remains before some other creature gets started on them.

Amphipods are a dominant life-form on tidal marshes. There are several genera and many species in each genus, all common from New England to Florida. Most amphipods look pretty much alike except to people who are familiar with their anatomy. For our purposes there are scuds, beach fleas, sand fleas, beach hoppers or seaweed hoppers and walking sticks. Males of some of these tidemarsh amphipods have among their appendages a pair of grasp-ing claws. Guess what they are for.

One example of the difficulty you will have discrim-inating between species is to be found when you compare *Orchestia grillus* and *Orchestia uhleri*. Count the segments of the flagellum on the second antennae. *O. grillus* has 20 or more segments, and *O. uhleri* has 12. Another species of *Orchestia* is *O. agilis* (Fundy to Hatteras) and it is more apt to be found on beaches rather than tidemarshes. *O. palustris* is an olive green or brown tidemarsh species common almost throughout our range. Maximum length, one inch. Members of this genus have round eyes.

A second genus of amphipods is named *Gammarus* and the distinguishing feature is kidney shaped eyes. Worldwide, there are over 2,500 species. These are the largest organisms in the sub-order. Males of *G. locusta,* the scud, grow to a length of an inch and a half. They are found from Maine to North Carolina and color varies from red-dish-brown to olive-brown. Like the others, they swim on their sides or with their feet up but, unlike many of the oth-ers, when they are on land they lie on their sides but never-theless move rapidly. By way of comparison, shrimp swim with their feet down. *G. annulatus* is abundant from Maine to Connecticut, is found higher on the beach, is lighter co-lored and has dark bands with red spots on the sides of its abdomen.

Melita nitida, Massachusetts to New Jersey, and *Moera levis* are still other species whose names you may encoun-ter in the literature. Their natural history and appearance are similar. Since their gills are on their legs, they cannot stray too far from water. These gills must be moist in order that the amphipod can "breathe."

Female amphipods carry their eggs and young between several of their legs, in brood pouches. How they get through the New England winter has always been a mys-tery to me. Perhaps the eggs survive freezing and drying as do the daphnia. It is hard to imagine that a form of life which does not burrow or fly away or swim away can last out a winter on a New England marsh. Of course, in the

Carolinas, Georgia and Florida the situation is altogether different.

I believe that once a person knows the origin of a name, familiarity will replace mystery. Years ago I knew a lady named Enola and always thought she had an exotic and romantic handle. Later I learned that she was destined to be an only child and her parents simply reversed the spelling of alonE. So it may be with amphipods. Their legs - pods are used for both swimming *and* walking. *Amphi* - both. This is also the source of the name of the class of salamanders and frogs—*amphibians*. As you become acquainted with tidemarsh life, your first impression of Amphipods will be that here is a small "shrimp" which sometimes gets out of the water but swims on its side or with its feet up when it is immersed.

COPEPODS

Copepods are usually flea-sized, and frequently fleashaped crustaceans. There are probably more than 5000 species and many are similar. Cyclops is a typical copepod and you will find some species of them in your fine mesh net after a pass nearly anywhere in slow-moving tidemarsh waters.

Since they are so small they are frequently overlooked or ignored by tidemarsh visitors but for sheer numbers (and weight) they make up a good part of the total "biomass." Most are small enough to eat single-celled plankton and detritus and yet are large enough to justify being eaten individually by juvenile fish and also by the river herrings which feed by simply swimming with their mouths open.

THE MANTIS AND THE WALKING STICK

For want of a better place to mention and illustrate them, let us here record the names and pictures of two oddball tidemarsh creek arthropod creatures which may surprise you the first time you see them but really they are not rare. One is *Squilla*, the mantis "shrimp." When you first see it you will be reminded of a praying mantis. It grows to seven or so inches and is sometimes found in creek bottom sea lettuce, especially after a storm. The picture is that of

Squilla empusa, but there are about 200 other species (some a foot long) that closely resemble it. It is a mud burrower.

The next odd looking insect-like aquatic creatures you should be aware of are the suborder *Caprillidea*. I must mention again that I use these scientific names since many of these creatures have no common names and it is only with the scientific names that you can quickly enter the more detailed scholarly texts to identify a particular species. So, *Caprillidea* are commonly, to many of us, the walking sticks. They resemble garden variety walking stick insects and you will find them, an inch or less long, crawling over underwater vegetation. Technically, these walking sticks are amphipods and there are many similar species between Cape Cod and Virginia.

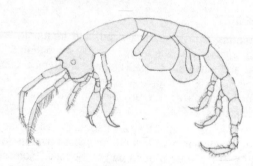

Walking Stick, *Caprillidea*

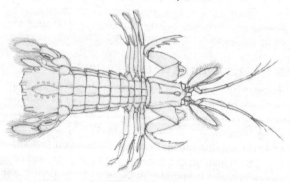

Mantis Shrimp, *Squilla*

FIDDLER CRABS

More so than any other creature, the fiddler crab goes with the Atlantic Coast tidal marsh. It would be hard to find a marsh without them. The pictures help if you are a first-time tidemarsh visitor. Afterwards, there is no mistaking them for anything else. They are burrowing crabs, they don't swim. The adult males have one tremendous claw. The other claw is the same size as both claws on the female. The large claw of the male can be either left or right. The males wave their large claws to defend territory and to fight other males. They also display these relatively enormous limbs to entice or seduce those of the opposite sex. Clicking of claws is also part of this display business.

Another aspect of their natural history is that they change color with the amount of light they are exposed to. They darken in sunlight and become pale in their burrows and at night. They have an internal clock synchronized with the low daylight tide and this is when they darken.

There are at least three species commonly found from Maine to Florida and the three prominent ones have similar habits. The small dark one you see in the grass is probably *Uca pugnax;* it is also known as the marsh or mud or black fiddler. Its burrows are found only in mud, peaty mud and grassy peaty mud. It is colored a uniform olive mud color when it is exposed to sunlight. This species tends to avoid sandy and high portions of the marsh. The male mud fiddler has a rough ridge on the inside of his large claw.

Uca pugilator males do not have the aforementioned rough ridge, and neither sex is uniformly colored olive mud, but they tend to have blotched multi-colored shells. This species is found higher on the marsh and more frequently in places where there is more sand and less mud. The burrows are likely to be in open area sand rather than grassy mud. When you see a "herd" of fiddlers on open sand anywhere from Cape Cod to Florida it is most likely to be not *U. pugnax* but *U. pugilator.*

The largest of the three prominent species is *Uca minax.* The male has the ridge on his claw but differs from *U. pugnax* in that there are red spots at the joints of this large claw.

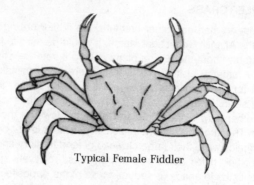

Typical Female Fiddler

Large Claw of the Male

Red Jointed Fiddler, *U. minax*

Sand Fiddler, *U. pugilator*

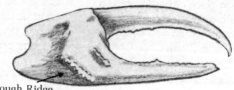

Rough Ridge

Mud Fiddler, *U. pugnax*

It is a lighter color than *U. pugnax* and the center of its top shell (carapace) is brown. This species might reach an inch and a half shell width in males where both of the others stop growing at about an inch width. *U. minax* tolerates, and even seems to favor, water which is nearly fresh and marshes which are less wet than those frequented by *U. pugnax* (mud) and *U. pugilator* (sand). *U. minax* is also known as the big fiddler and the red-jointed fiddler. One authority mentions that it can survive out of water for several days. *U. minax* males build an arch over the entrance to their burrows and they perch by this arch and display their mighty and their fearsome (and perhaps sexually attractive) large claws to passersby.

Fiddler crabs eat vegetable and animal debris and detritus which they find in mud. Although they have been observed eating under water, they usually eat on land. They differ from aquatic crabs in that they have not only gills but also primitive lungs which help them with respiration when the tide is out. Still another adaptation permits them to survive without the use of either gills or primitive lungs! It seems they are able to withdraw, by chemical means, oxygen from stored glycogen (a product of their metabolic process). The glycogen breaks down to make available some oxygen, which is then consumed in the course of living, and lactic acid, which is stored until oxygen is again available. Then, the lactic acid is re-oxidized and it again becomes glycogen. All this in a little fiddler crab just an inch across, when it is fully grown, and too minute to see with the naked eye when it is hatched.

Fiddler crabs are relished by all birds, mammals, fish and even other crabs who can capture them. There is no need to list them, their name is legion.

The fiddler eaters, their name is legion, but still the fiddlers survive. They are with us because they are prolific, also they survive because their burrows are not easy for predators to penetrate. The burrows are barely large enough to pass their occupants. Generally each burrow averaging one foot in depth was built by its owner who occupies it alone. The burrow commences in an acute angle with the ground and then after a few inches it generally turns downward. In the case of a male, it ends in a chamber where he hopes to entice a female, there to mate with her. A female's burrow does not terminate in such a large chamber but it may be connected by a side tunnel to the chamber of a nearby male. She can visit him, if she desires, without surfacing, but he cannot get into her burrow since he is always larger, especially with that great fiddle of a claw. People with a ribald sense of humor can have fun with this.

The fertilized female extrudes her eggs into her abdominal flap—as many as a quarter of a million eggs in one small spongy cluster; and when they are about to hatch after being carried for approximately two months, the female releases her brood into the nearest creek at high tide. The young fiddlers pass through several moults (with metamorphosis) in the water before they return to land for the remainder of their lives.

They also survive because under stress they can abandon a limb, even the great claw of the male, to an enemy or predator and in the confusion of the sudden parting, they frequently escape to regenerate the limb at the subsequent moult.

Northern fiddlers hibernate; in the far South they are active throughout the year so long as the temperature is right. The minimum temperature for activity by *U. minax* is 68° F. Both *U. pugilator* and *U. pugnax* tolerate lower temperatures; they begin to stir at 60° F. If the noonday sun overheats a fiddler, or if a cool breeze should make it feel uncomfortably chilled, the little crab will retreat to its burrow until conditions are more to its liking.

MARSH CRABS

Another crab often associated with fiddlers is *Sesarma reticulatum,* the marsh crab. It looks somewhat like a female fiddler with both claws the same size. It burrows in the same manner as the fiddler. To identify it in fiddler country note the shape of the top shell (carapace). Fiddler carapaces are wider in front than in the rear but marsh crab carapaces are rectangular. Fiddler females have smaller claws than do either sex of this species. The marsh crab is common everywhere south from Cape Cod.

MUD CRABS

When you find a fiddler-sized or larger crab in fiddler country with two unusually thick heavy claws of nearly equal size and a slow way about it you are probably looking at a mud crab. There is a family of these creatures with genera and species distributed all over our East Coast tidemarshes. Since they are frequently found with or near oysters and oysters are frequently transplanted, it is entirely reasonable to expect to find specimens out of their natural range.

Most of these crabs bear generic names which include *Panopeus*. For example, there is *Eurypanopeus deprossus* from Long Island Sound, south and *Rhithropanopeus harrisii* everywhere in brackish tidewaters.

Mud crabs tend to be more aquatic than fiddlers and they often eat fiddlers when they are able to capture them.

Marsh Crab, *Sesarma*

Mud Crab, *Panopeus*

Green Crab, *Carcinides*

GREEN CRAB

This species, *Carcinides maenas* is easy to recognize. You will find it from Virginia, north. The largest sports a shell less than three inches in its greatest dimension. It does not have the two points on its shell which make the blue crab so easy to identify nor does it have swimmerettes on its last pair of legs. The same-size large claws tell you it is not a fiddler and the deeply sculptured "teeth" in the shell are distinctive. This crab is sometimes a tidemarsh creek dweller but given a choice, it tends to rocky areas of seawater salinity. I did not mention the color of the green crab for identification purposes since many are not green but rather a nondescript rusty brown. Green crabs are considered a good bait for blackfish (tautogs).

BLUE CRAB

The blue crab is wonderfully adjusted to its life in the typical tidal marsh. It is aquatic but will survive out of water for hours if not sunbaked or frozen. It eats all manner of things from debris to live fish. It burrows in the mud of creek bottoms or skulks within caverns hollowed in the peat of creek banks. It swims rapidly. It adjusts instantly to salinity changes from fresh water to sea water. It is active day and night if the temperature is sufficiently warm; and it becomes dormant when the water is cold and food is scarce. The female carries her eggs with her until the napulii hatch, thus assuring that as long as she lives, her eggs are safe from predators. Most people who have studied blue crabs, or pursued them will agree they are intelligent, strong, fierce, feisty, fast, graceful and delicious.

The scientific name of the blue crab, sometimes also called the blue claw or common blue crab or edible crab or blue swimming crab, is *Callinectes sapidus.* The word *callinectes* means beautiful swimmer and *sapidus* means palatable or having an agreeable flavor. No doubt about it. Similar species are *C. hastatus* and *C. ornatus.*

Blue crabs are found from Cape Cod south. The greatest concentration seems to be in Chesapeake Bay where an industry and a whole way of life have evolved around this decapod crustacean arthropod. Scholarly texts refer to crabs as Decapoda because they have ten feet. First let's look at the life history of the one that got away. An adult male can inseminate more than one female, but an adult female completes its life's mission with one cluster of eggs. When they are newly accumulated under her belly flap (technically this is the "abdomen, short and reflexed"), the eggs are in a spongy mass, orange when new, progressively browner and darker as time wears on. Then the eggs hatch, still attached to the abdomen of the female and the young are set free. They then look like tiny insects, they have tails, prominent eyes and little resemblance to adult crabs. All species of crabs go through two stages of metamorphosis as they moult after they become free-swimming.

Growth requires moulting since the shell does not grow

and hence it limits the size of the organism. A really large blue crab might approach nine inches measured from point to point across its shell. Females are impregnated while they are soft from having only recently moulted; the male who found a soft shelled female would necessarily be a "hard shell" at that particular time. She carries his sperm until she is ready to release her eggs from inside her body cavity to the external folded flap which is her abdomen. The timing is a function of water temperature and varies according to location, but regardless, an egg cluster usually ushers in the female's last official act. By the time she has performed this task and her eggs have hatched, she is perhaps two years old and will soon die of old age. The male similarly is believed to live less than three years, whether he breeds or not.

In tidal marshes, the sexes seem to segregate, except of course when they are mating. In winter the females hibernate in deeper waters which are perhaps more saline. The males seem to favor the higher and less saline creeks.

Blue crabs are scavengers and are therefore several steps removed from the source of a food chain. This makes them especially vulnerable to pesticides and similar poisons, and the chances for their survival seem to parallel that of the osprey. For example, both declined and later showed signs of improvement in Long Island Sound at about the same time.

To catch blue crabs on tidal marshes, one must be smart, patient, quiet, and quick. In most places laws prohibit us from taking them while they are hibernating. To capture them then where it is legal, dredges are drawn through mud by boats traveling at slow speed.

When blue crabs are active, watermen find that eel is an excellent bait, but recently the price of eels has gone so high as to drive them out of the bait category. Actually, blue crabs will eat almost anything from the frames of filleted fish to chicken necks. Two crabbers in a small boat can do nicely without bait if one paddles, poles, or rows against the current (to avoid the obscuring smoke-screen of mud), and the other, in the bow, with a long-handled net (the handle is usually four and a half or five feet long), dips them up as he

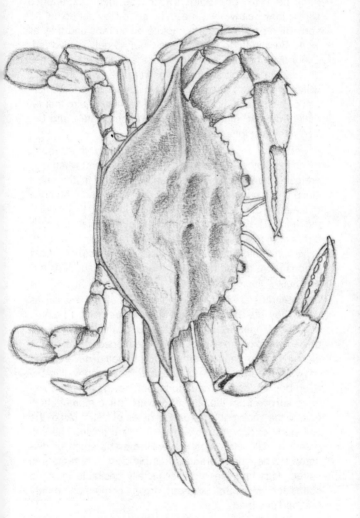

Blue Crab, *Callinectes*

sees (or senses) them hiding in creek bottom mud or creek bank peat. No one bothers to chase them; experience shows they nearly always escape once they get going. All sorts of traps are made to capture blue crabs and they all work. Be careful to know the law before getting involved. Every state has its own regulations and these sometimes change at the arbitrary line demarking the sea from inland water. Other laws govern commercial versus noncommercial crabbing and still others in most states require that females with egg masses be released immediately and unhurt whenever they are taken.

Soft crabs never take bait but they are sometimes found by netters. Others are about to moult (shedders) when they are caught and these are frequently held until that moment since a soft crab is an especially desirable dish. The color of the edge of the swimming legs is an index to the oncoming moult. The crab which is about to pop its shell develops a milky swimming paddle with a bright red outline. Incidentally, concerning color, the top shell (carapace) of a blue crab is more green than blue. It is the claws and legs that are mostly blue.

Blue crabs are fast in the water and out of the water as well. They are extremely strong and can inflict a painful, perhaps dangerous bite. Handle them with a net or with dexterity and great respect. If you are nipped, break off the claw instantly; chances are the crab would never let go anyway. Then break the claw apart. They keep well in dry deep buckets or damp burlap bags. They do not need water to survive from the time of capture to the moment they go into the cook pot for twenty minutes of steam. All of the soft parts except the pale gray gills are eaten. The soft greenish material is mostly liver. It may be eaten or discarded. The bright orange crumbly part in a female is stored eggs and these are especially desirable in some soups and stews. Soft crabs are broiled or more commonly breaded and fried.

If you are interested in blue crabs, you owe it to yourself to read William Warner's *"Beautiful Swimmers,"* Atlantic Monthly, Little, Brown and Co., 1976. It is by far the best book on the subject I have ever read or heard about.

HERMIT CRABS

Several species of North Atlantic crabs have evolved so that their abdomens are unprotected by the usual hard crab shell and further, their abdomens are twisted in the same direction as the twist of most North Atlantic snail shells which they occupy; and this abdominal twist also is as the result of evolutionary development.

So the hermit crab has traded away the protection of his ancestors' shell for the protection of a snail's shell. He has gained the protection of the really hard gastropod shell but he must give it up as he grows out of it, or if a stronger hermit crab fancies that particular snail shell and desires to occupy it. Lacking the perspective of a hermit crab, I question the value of the trade-off; but in spite of my skepticism, hermit crabs are common on the entire length of our U.S. East Coast. You will surely find a few hermits in the shells of mud snails *Nassarius,* (also sometimes known as *Ilyanassa*) and periwinkles (*Littorina*) in the tidemarsh creeks.

If you wish to see an entire live hermit crab without hurting it, you will have to get behind it. Drill or grind a tiny hole one and a half snail twists back from the entrance apperture and insert a broom straw or a piece of monofilament fishing line in the opening you made. A little twisting and gentle pushing will tickle the crab out of his snailshell home. Leave the homeless crab undisturbed in an aquarium with some snail shells and it will quickly find its old home and repossess it. Incidentally, hermit crabs do not kill snails to acquire their homes. The shells are invariably those of snails that died of other causes.

Hermit crabs are scavengers, in common with most other crabs, and they do very well in captivity where they continue to perform this useful function. Choosing and gaining and retaining occupancy of its successive homes is an important thing in the life of a hermit crab. Whether you consider this activity as "antics" or "animal behavior" depends on your point of view. Some people find that hermit crabs cooked in garlic flavored oil are a delicacy.

The species of hermits you are most apt to encounter in a tidemarsh creek are the warty large hermit crab *Pagurus pollicaris* which grows to five inches long and frequently occupies moon shells or whelks when it reaches its maximum size. Its broad claws are rough and hairy. The other species is *Pagurus longicarpus* which never gets much more than three inches long. This latter species has relatively, long, slender, virtually hairless legs and claws.

BARNACLES

The typical barnacle is the only *fixed*, really fixed, crustacean. Believe-It-or-Not Robert Ripley made real money telling people that barnacles are crustaceans and not mollusks, and he was right. Young barnacles are free-swimming but soon they cement themselves to something solid and there remain fixed for life. They are found on rocks and driftwood and piling and on crabs, and snails, and mussels, and oysters, and even other barnacles. They attach to ship bottoms in such great numbers as to slow the vessel. One species attaches itself to whales, and another is a parasite to green crabs. Conventional barnacles are beautiful to watch as they wave the feathery bristles on their "legs." They are awful to step on when one is barefoot. Large barnacles are eaten by people and smaller ones are mostly left

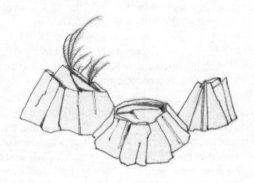

Barnacles

alone. Barnacles eat phytoplankton and zooplankton only when they are underwater. If one happens to attach itself to something fixed above the low water mark, then there will be times when it must close up shop and await the new incoming tide.

In keeping with the general disclaimer which appears elsewhere in this book, the barnacle provides an exception to the crustacean "rule" of moulting its shell to provide room for growth. Here, it seems, the tergas and scutas enlarge at their edges to cover the soft parts as growth continues.

The common barnacle from Delaware Bay to the north is *Balanus balanoides,* the rock barnacle. In this species the attached undersurface is a membrane. It favors high salinity and is not found in the upper reaches of tidal creeks fed by fresh water. Its low salinity relative is *B. eburneus,* the ivory barnacle. These ivory barnacles are sometimes found in salinities so low you can drink the water. Here the undersurface is as hard as the tergas. The 'door like' plates which cover the soft parts of barnacles are called tergas and scutas. The configuration of these tergas and scutas helps us determine the species, but barnacles themselves certainly don't use our method. If they have any vision at all, it is certainly not that acute. Reproduction among *Balanus* is accomplished by close barnacles who reach across to each other and transfer sperm through tubes. If this is the only way they do it, some individuals are destined to a lifetime of celibacy. I don't know, but I suspect there is more to it than that.

What we do know is that barnacles are hermaphrodites. That is, each one can function as a producer of both sperm and of eggs. It is unlikely that for most species, any particular barnacle would routinely produce the sperm that fertilized its own eggs since when it is doing one thing it is not apt to be also doing the other. For one interesting exception, read on.

SACCULINA

Our ordinary tidemarsh green crab *Carcinides maenas* is parasitized by a species of barnacle in the most bizarre manner; it is to me the strangest example of parasitism in the animal kingdom. (MacGinitie describes it in some detail.) Briefly, *Sacculina*, the parasitic barnacle, produces eggs which hatch into napulii and swim freely and eventually metamorphose into a cypris form. The larval cypris *Sacculina* swims until either it perishes or it finds a green crab. It attaches itself to a "hairy" process on the crab and injects part of its cellular structure into the crab through the hair. Eventually these cells get into the bloodstream of the crab and come to rest in the abdomen. You can forget the part that remained outside, it has completed its function and is finished. But the cells that got inside grow and send "roots" to all parts of the crab's body. A little like a cancer, if you catch the pun. These roots take nourishment from the crab and also they completely absorb all the crab's reproductive organs. By the time the crab moults, the parasite is on the inside and is not disposed of in the shedding process. With the crab's reproductive organs absorbed and replaced by another organism, it really isn't a crab anymore. That is to say—that thing doesn't lay crab eggs, it lays *Sacculina* eggs. This parasite is a perfect hermaphrodite, it fertilizes its own eggs and looks for all the world like a swollen female green crab which it has taken over, or taken into. If perchance the little microscopic cypris form of the *Sacculina* landed on a male green crab, the crab would soon lose his gender and he would appear after his next moult like a female which he would not be, because really all he would be is the shape of a female green crab with the reproductive capacity of a hermaphrodite barnacle. If the *Sacculina* is removed from what had been a male green crab, he will, when he moults, become a hermaphrodite green crab. So what else is new on the tidemarsh?

SHRIMP AND PRAWNS

A short review of the scientific names is appropriate here since you will need it to go into the more specialized textbooks for positive identification of the many tidewater species and furthermore, many do not have common names.

Shrimp are in the:

> Phylum *Arthropoda*-with jointed legs; including insects

> Class *Crustacea*-gill breathing, segmented and bearing two pairs of antenna; including forms with all numbers of legs

> Order *Decapoda*-ten legs attached to the thorax; including crabs

> Suborder *Macrura*-with large abdomens which terminate in tail fans; including lobsters and crawfish

> Section *Natantia*-free swimming shrimp

Now, finally, among the section *Natantia* we find various tribes and families and genera and species of shrimps and prawns. Hundreds of them. Between Maine and Georgia you might encounter a hundred species in a lifetime of intensive collecting. But except for the more scholarly purists we can do some lumping and thus help to identify the species of shrimps you are most likely to encounter.

You may need a hand magnifier; not a microscope but just a four to seven power magnifying lens. Look at a shrimp-like animal and confirm that it has ten legs (*Decapoda*) and a tail fan (*Macrura*). It is probably an isopod or cumacean or amphipod if there are more than ten walking legs. Is it a creeper or a swimmer? Creepers are the crayfish and lobsters both found only infrequently in tidemarsh waters. Swimmers are *Natantia*. Now look at those *ten* legs. Do they have any claws on them? Which ones have claws? This should narrow it down to the families and then color and the shape of the spine between the eyes should be compared to the illustrations in the detailed field guides. Don't invest any sincere money on the identification of shrimp—it is as chancy for beginners as playing cards with strangers on the train.

All right, let's get on. The most common tidemarsh creek shrimp is *Palaemonetes vulgaris*, the common prawn. It has small claws on its first pair of legs and larger claws on its second pair of legs. This species is found in saltmarsh waters from New Hampshire, south. Maximum size is one inch, not including the antenna. It is greenish and some-

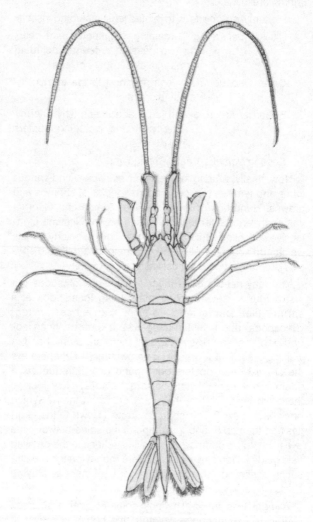

Sand Shrimp

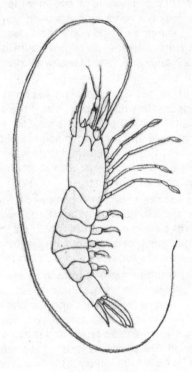

Peneus setiferus

what transparent. The common prawn spends its entire life in tidemarsh waters.

Still another species that will show up when you seine is *Cragnon vulgaris*, the sand shrimp. *Cragnon,* like *Palaemonetes* has two pairs of pinching claws but *Cragnon's* claws on its first pair of legs are especially large. The sand shrimp grows to a maximum length of two inches. Its color is a nondescript speckled gray. This species is also known as *Crago septemspinosus*. It is abundant from North Carolina, north. Several related species resemble it closely and detailed examination is required to identify them.

A tribe of shrimps with grasping claws on the first three pairs of legs are *Peneidae*, and the first of two species you are most apt to find is *Peneus setiferus*. This is one of six species that make up the common shrimps of commerce. It is found from Virginia, south, and large individuals (over five inches) are called prawns. Smaller prawns are obviously called shrimps; obviously.

Another species of common commercial shrimp is *P. braziliensis* and it is found from Cape Cod, south. The dorsal keel on its carapace runs back only 2/3 of the total carapace length. Adults of this genera are seagoing (pelagic), but the juveniles use tidemarsh waters as a nursery.

And yet another small common grass shrimp is *Hippolyte (Virbius) zostericola*, the eelgrass shrimp. It is found between New Jersey and Cape Cod and superficially it resembles the other grass shrimp, (P. vulgaris) except that it is mottled green and brown and is sometimes peppered with red. The abdomen (mostly the meaty part—the part we eat) is bent down sharply at the third segment.

Let's wind up this consideration of shrimps in the manner of the successful old time preacher who explained "I tells 'em what I'se gonna tell 'em and then I tells 'em, then I tells 'em what I told 'em." All right. Shrimp have long thoraxes and long abdomens with fanlike tails. Ten legs, some perhaps with claws, are attached to the thorax. They swim with their legs down. Amphipods swim on their sides, or with legs up and they have more than five pairs of legs.

You may discover that all the shrimp you find of one

particular species are the same sex or are juveniles. This should come as no surprise since some species of shrimps start out as males and become females only in adulthood. Also remember that some species are seagoing as adults and only the juveniles inhabit the tidemarsh creeks and bays.

When it comes to food chains, shrimp are pivotal. Some eat plants and others dine on animal life, and still others are scavengers. In turn, they are eaten by each other, and virtually every other predatory creature.

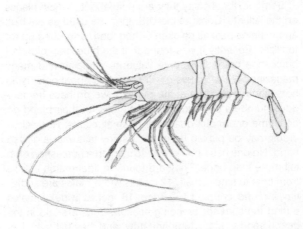

Common Prawn

HORSESHOE CRAB

The marine arachnid with the sword is, to look at it, a monster from prehistory. Fossil records actually show it to be very similar, but not identical, to the trilobites of Cambrian times. Genuine modern fossil horseshoe crabs occur for the first time much later, in Silurian deposits. Although we commonly call it a crab, it is not a crab; in fact, it is not even a crustacean (crab, lobster, shrimp and barnacle). *Limulus polyphemus*, horseshoe crab (also called the king crab on our Pacific coast) is most nearly related to the spiders (arachnids). Unfortunately, and perhaps adding to the confusion for unscientific readers, there is also a "spider" crab of the northern Pacific which is not an arachnid. The spider crab is a true *Decapod* crustacean and it too is sometimes called the "king crab"; some of them have been known to span over eight feet!

Some textbooks say they are edible, but in most places on the Atlantic Coast of the U.S. they are used as eel bait, and in times past as chicken or hog food, or ground up for fertilizer. Recently it was found that the blue-gray blood of horseshoe crabs has some value in medicine and many magazine feature writers got very excited about preserving horseshoe crabs; but at the moment we do have tremendous numbers and those we don't milk for their blood do consume enormous quantities of small clams. Horseshoe crabs may be picked up by their spiny tails with a gloved hand. No one has been known to step on a horseshoe crab tail more than once. They are found in tidal marsh creeks from time to time. Small (12 inches wide) males are sometimes found coupled to larger (18 inches wide) females during the summer breeding season. Eggs are laid in the beach sand in May, June and July, and they hatch in July and August. The egg laying and the hatching are regulated by spring tides. In the Far East, these eggs are considered to be a delicacy.

ARTHROPODS - INSECTS

The typical hexapod insect has a head, a thorax and an abdomen. There are three pairs of legs and frequently one or two pairs of wings. It breathes air as an adult. Some are predators; others scavenge, suck plant juices, eat vegetation; and still others are parasitic. Some creep, some fly and others live mostly in or on the water. There are, on earth today, more species of insects than all other life forms combined; supposedly 75% of all known species of animal life are insects. The words insect and hexapod are synonymous. The name hexapod is derived from the Greek *hex* = six, and *podos* = foot. Grasshoppers, crickets, beetles, mosquitoes and flies are typical insects.

GRASSHOPPERS AND CRICKETS

Since saltmarsh hay is no longer a universally important agricultural crop, there is now plenty of tidemarsh vegetable matter for insects to suck and graze upon, and there are plenty of insects that do just that. The dominant grazer of the various saltmarsh hays is the ground cricket *Nemobious* species. These crickets seldom exceed one-half inch in length and they are brown or velvety-black all over. And there are tremendous numbers of them. The same can be said for the meadow grasshopper *Conocephalus* species. Most are green with red-brown striping on their heads. Others are brown. They are all smaller than three-quarters of an inch long but what they miss in size they make up for in numbers.

After an adult female cricket or grasshopper has been fertilized, she deposits her eggs in a cavity which she excavates in the ground and the eggs hatch into miniatures of the adult which then moult from time to time, since their hard outer covering is incapable of growth. These insects are capable of hibernation, and certainly the eggs which are laid in late summer remain dormant over the winter and then hatch the following spring. The number of adults that survive the winter depends on the latitude. It is unlikely that Carolina grasshoppers and crickets hibernate, perhaps they are less active on the coldest days. It is also unlikely that large numbers of adults survive a New England winter, no matter how hard they try.

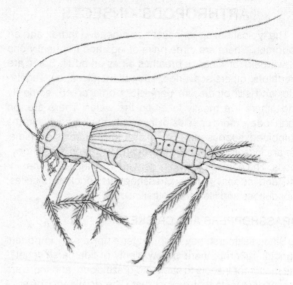

Ground Cricket

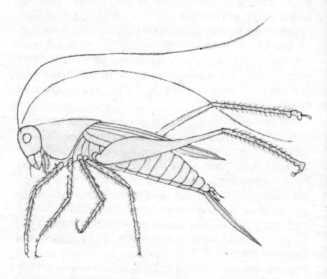

Meadow Grasshopper

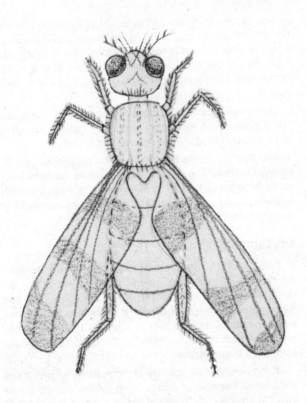

Picture-winged Fly

FLIES

There are three forms of tidemarsh flies. Two are voracious predators both in larval form and as adults. The smaller of them is often called the deerfly; it is black or brownish black with dark markings on the wings. It is approximately the size of a housefly.

The large fly of the same family (*Tabanidae*) is the horsefly or green-head. Its eyes are green and it will suck blood from any warm-blooded creature. This is the fly that took the joy out of cutting salt hay on the tidemarshes. Horses and oxen were sometimes driven into a frenzy from its bites.

Even the larvae of both these flies are fierce. The grubs live in marsh mud and eat anything they can kill, including isopods, amphipods, other insects, and worms.

The third fly of the tidemarsh is the picture-winged fly (family *Otitidae*). The wings are banded and the overall color is brown or blackish brown. These are smaller than houseflies and do not suck blood, but rather the juices of plants which they seem to sponge up.

MOSQUITOES

Often when I visit a tidemarsh in summer, I wonder how the millions of mosquitoes found sustenance before I arrived, or how they will manage after I am gone. One can easily get subjective about these gnat-like insects. The very existence of tidemarshes in or near settled areas has sometimes been decided in terms of the discomfort and disease caused by these relatives of the flies.

Of the nearly 2000 species of mosquitoes known, there are several capable of reproducing themselves on saline tidal marshes and their natural history is neatly tailored to fit their environment.

Let's begin subjectively, with the one that just bit you. It was a female. We may not readily identify the species, but common to all mosquitoes, only the females suck blood when they can get it. Adult males drink plant juices throughout their winged lives and the females also can survive on this diet but they should have one blood meal to properly ripen their eggs. It is possible that several generations can be produced without blood, but this would be the exception, rather than the rule.

All mosquito eggs are deposited in water or on land in anticipation of water soon to come. A female may (depending on species) lay from 40 to 300 eggs singly or in clusters. When the eggs have been wet and the triggering mechanism is released, the larvae hatch and they breathe air from the surface of the water. Most of these wrigglers eat vegetable matter, some species are predatory and even cannibalistic. Technically, the eggs hatch to larvae and these straight shaped larvae grow and moult and eventually after several stages metamorphose into another water form called pupae which are somewhat curved or crescent shaped. The pupae then metamorphose into the winged form which eventually mates and sends its genes into the next generation.

The tidemarsh forms start with eggs laid in mud above the normal highwater mark. A series of springs or perigees or spring perigees will flood the area and it is then that the water cycle is completed and the winged stage emerges. This is a fortnightly thing and can be anticipated with little glee except by the swallows, martins, bats and dragonflies that feed on the flying forms and the black ducks and mallards that eat the larvae and pupae.

Aside from the itch that accompanies the bite is the problem of disease transmitted by mosquitoes. Malaria is the classic, but yellow fever, elephantiasis, encephalitis and heartworm disease in dogs rely on mosquitoes as vectors between one warm blooded animal and another.

Control by ditching is proven by now not to be the complete answer, chemical poisons are not selective enough to satisfy all our environmental concerns, and oil also destroys not only the mosquito but many valuable life forms as well. Chemicals which destroy the surface tension of the water and make it impossible for the larva to breathe have been used experimentally, but it seems to me the best control at the tidemarsh is by fish such as *Fundulus* and *Gambusia*, and the best control near habitations is by fogging the air periodically with non-persistent insecticides and the local application of repellants on skin and clothing. Probably, if I ever had encephalitis or malaria, I would change my tune.

That triggering mechanism which I mentioned earlier may be set off one or as many as five or ten years after the eggs are laid, thus assuring continuity even if there are several years of drought or other unfavorable conditions. Freezing is handled by mosquitoes in two ways. Some species lay eggs at the summer's end which don't hatch until the following spring and with others, the female hibernates and lays her eggs only after she is awakened by the balmy breezes.

One common saltmarsh mosquito of the northern states is *Aedes sollicitans*. It is brown with distinct white bands on its beak, legs and abdomen. Another white band appears lengthwise on its back. As mosquitoes come, it is medium sized and when with its downturned head it bites you, its body is parallel to the surface of your skin. By contrast, the malaria mosquito (*Anopheles*) bites with her head down, her proboscis is in a straight line with her body.

Since a mosquito might travel several miles from its hatching place before being sighted, it is impossible here to anticipate what other species beside *A. sollicitans* will show up on a tide marsh.

THE CHORDATES - BEGINNING WITH THE FISHES

Generally speaking this is the phylum of animals with dorsally located nervous systems. We are mostly interested here in the typical backbone animals. Actually the backbone animals *Vertebrata*, are but a subphylum and on the tidemarshes within this subphylum there are the familiar classes of:

    Pisces (fishes - authorities today break down the class of *Pisces* into two parts, the cartilagineous and the bony.)

    Amphibia (salamanders, toads and frogs)

    Reptilia (snakes, turtles, lizards and crocodilians)

    Aves (birds)

    Mammalia (mammals)

This business of formal classification is a little tedious when you are not accustomed to it; but as mother used to say about spinach, "You may not like it now, but when you grow up big and strong, you will thank me for making you eat it."

Fish Topography

Spiny Dorsal · Soft Dorsal · Adipose Fin · Caudal Fin · Operculum · Lateral Line · Barbel · Pectoral Fin · Spiny Ray · Ventral Fin · Anal Fin · Finlets

PISCES, THE FISHES

Pisces are cold-blooded aquatic chordates—hairless and featherless, but some do have primitive feet. There is not much more I can say about fish without finding some exception. Some are scaleless and others are not even slimy. Some have lungs and can live out of water for extended periods. Some lay eggs and others give birth to living young. Like snakes, all fish lack eyelids.

Tidal marsh waters support a score of species that are rarely found elsewhere and they are the species this book is primarily about. I will not dwell at any great length on visiting non-residents. The full-time residents have one thing in common. They are able to tolerate salinity changes with no inconvenience. There isn't any pattern to their lifestyles. The eel, for example, is a scavenger that was hatched in the Sargasso Sea and spends part of his "growing up" in tidal marshes. Yes, *his* "growing up." The female eel remains in fresh water for some time between two and twelve years of growth but the male seems to choose to grow up in brackish tidemarsh waters and estuaries. You see, there is a difference. By contrast, the white perch and striped bass spawn in less saline waters than they frequent as adults for added contrast, the *Fundulus* (killies) are permanent residents of tidemarsh waters and they live out their entire life cycle within a mile of their birthplace.

KILLIES

Think of a small active stout-bodied fish in a small, disorganized school in a brackish tidal marsh creek, and you are probably thinking of a member of the genus *Fundulus*. Technically, it will probably be *Fundulus heteroclitus*—the common mummychog or killie; *F. majalis*—the striped killifish; *Cyprinodon varigatus*—the stumpy little sheepshead minnow; or *F. chrysotus*—the golden ear of the southern states.

Kids catch killies in milk bottles baited with bread. Fishermen do likewise with minnow traps baited with bread, crushed mussels or crushed crabs. Small seines, usually eight feet long and 24 or 30 inches deep with a pole at each end are also effectively used. When seining for mummy-chogs you should expect to get some grass shrimp, small perch, eels, some striped killies and a few sheepshead minnows as well. All these species of killies make great bait for fluke (summer flounder) and they also can be used in a pinch as blue crab bait. Small mummychogs—also spelled mummychugs—are good bait for large white perch, but most perch fishermen prefer to use an artificial lure or a worm.

F. heteroclitus is capable of living in all levels of salinity from zero to thirty-five parts per thousand and in about all temperatures from 30° F. to 85° F. The other species are found less frequently in the full range of salinities. It is truly a tidal marsh fish for all seasons. It tolerates uncomfortably warm waters and becomes dormant in the mud when the water temperature approaches freezing. Mummychogs eat just about everything that is small enough to swallow or can be torn apart, and they in turn are eaten by just about everything that is larger in similar manner. Mullet have the same body shape as killifish but all species of mullet have two dorsal fins, while all species of killifish have but one dorsal fin. Also mullet are bottom feeders and their mouths are located appropriately lower. The largest mummychog is six inches long, but these are rare and most do not exceed four inches. Their general overall appearance is more like sunlit mud and grass than anything else.

The mummychog has adapted its reproduction to the tidal cycle in a wonderful way. It is not unique among fishes but wonderful nevertheless. It sometimes deliberately "strands" its eggs so that they are—most of the time—out of water, by depositing them high up on the leaves of *Spartina alterniflora*. Spawning takes place near the surface during nocturnal new moon or full moon spring tides in Delaware Bay. This is reported in a study by Taylor, Di-Michele and Leach; *Copeia,* 1977, Number 2 pages 397-399. This article and the literature cited by these authors invites a lifetime of additional study. Briefly, the authors of the article observed that during spring tides, the summer daytime high tides are lower than comparable night highs on the Atlantic Coast of North America. The difference in Delaware Bay was a matter of eleven centimeters (4.3

inches) and it was only in this part of the *Spartina* plant that the mummychogs attached their eggs to the inner leaves.

The eggs, therefore, were surely wet only during the top of the tide at night for the first few days of the spring tide during which they were laid and subsequently only barely wet for several days during the neap. The next spring tide would give the eggs a good soaking and they would then immediately hatch and the young fish would find high tide pools for their early development. Here they would be relatively safe from many predators and safe also from strong tidal currents that might sweep weak swimmers out to sea. In this particular instance no eggs were found in those areas always under water.

The same species of fish is known to spawn in Connecticut in tangles of filamentous algae and in Virginia inside empty horse-mussel shells. In all these situations, mummychogs seem to have selected sites for their eggs which provide protection from both aquatic predators and excessive drying. This is a problem in compromises which works well, the mummychog is an abundant species from Florida to the Gulf of St. Lawrence.

Fundulus majalis, the striped killifish grows much larger than the mummychog. I have even caught them on baited hooks. The shape is the same but the total length might reach eight inches and even larger specimens have been found near the warm water discharges of electric power generating stations. Young striped killies have about nine dark narrow vertical stripes between the head and the base of the tail. Adult males have a black spot on the rear of the dorsal fin and three or four more of these stripes, but adult females have two to four horizontal stripes and perhaps just one or two vertical stripes near the tail. *F. majalis* is found from Florida to New Hampshire in brackish water of varying salinity. Generally they favor saltier waters than do their smaller relative *F. heteroclitus.*

The only really stumpy killifish you will encounter on tidemarshes is commonly called the sheepshead minnow. Scientists named it *Cyprinodon variegatus* and its name suggests that it is decorated with a variety of colors. This is true. It is mostly olive but yellowish below with some suggestions of irregular dark vertical bars and during the

Fundulus heteroclitus, Mummychog

Juvenile Male

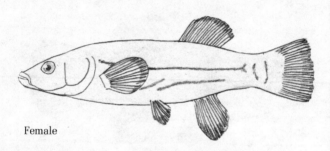

Female

Striped Killie, *Fundulus majalis*

Sheepshead Minnow, *Cyprinodon*

breeding season the males are somewhat bluish above and orange below. The female sports a dark spot on the rear part of her dorsal fin.

These three-inch fish are mostly vegetarian but nevertheless pugnacious. The eggs are laid in clusters held together with mucous threads and the eggs hatch in five or six days. Sheepshead minnows are found in tidemarsh waters from Maine to Florida.

The golden-ear, *Fundulus chrysotus* is shaped like a mummychog but is much prettier. In fact it is sold in pet-shops and kept by aquarists in fresh, brackish and even nearly 100% saltwater tanks. These three-inch fish are pale olive with blue and gold scales and scattered small red dots especially on the males. In addition some males are flecked with specks of black—a genetic trait known as melanism. Golden-ears will eat some vegetable matter but they are more properly thought of as pugnacious predators. They eat small crustacea and insect larvae avidly, and anything too large to swallow will be quickly torn apart and then consumed. This species is native to South Carolina, Georgia and Florida.

There are still other killies to be found in our tidemarsh waters but positive identification of these species requires fin ray counts or other precise measurements. These include *Fundulus diaphanus,* the banded killifish which is found between Quebec and South Carolina. Males sport 18 to 22 dark vertical bars and the females have 14 to 16. But for the bars, it looks much like a mummychog.

Still another in this series of killies is *Fundulus luciae* the spotfin. It too, looks like a mummychog but the male has a dark spot near the back of his dorsal fin. These fish are found from Connecticut to Georgia in brackish water.

If your killifish sports a single horizontal line and no vertical bars, it might be *Lucania parva,* the rainwater killifish. This species is found in all levels of salinity between Cape Cod and Texas. Maximum size is three inches and the only oustanding feature is that it is the only common tidewater killifish with a single horizontal stripe and no bars. The male has a dark area on his dorsal fin.

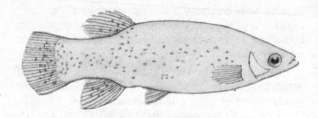

Golden Ear, *F chrysotus*

MULLET

In Florida, mullet are called lisa in an attempt by fishermen to get more people to eat them, but regardless, they are a commercially important fish. The striped mullet is found from Long Island Sound, south. This is *Mugil cephalus*, and it grows to a length of two feet. In cold waters it is said to hibernate in the mud from whence it derives its food. Juveniles eat crustacea but adults seem to subsist on plant detritus found in the mud. The mullets have two dorsals, forked tails, round bodies and are frequently found in tidemarsh mud-bottom creeks. The mouth is set low, which suggests a bottom-feeder.

The other species ranges from Cape Cod, south and is commonly also called white mullet (or lisa). Technically, *Mugil curema*, it grows to three feet and looks much like *M. cephalus* except that adults lack the horizontal stripes.

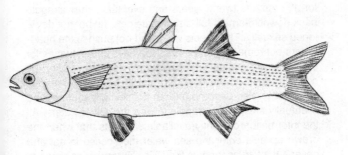

Mullet, *Mugil*

EEL

The natural history of the eel is a romance as rich in detail and inexorable drive and in as yet unexplained behavior as that of the indefatigable Atlantic salmon or the timely swallows at Capistrano. But until now there have been no epic poems, no songs, no lasting memorials to this irrevocable constancy of the eel. When viewed by poets and other people-oriented people, salmon and swallows probably look smarter and more determined and this may be the anthropomorphic determinant to their fame. So, let's set the matter straight. Permit me to start with a two-inch blackish pencil-lead of a fish swimming in a school with tens of thousands like it upstream from a river mouth.

If it is the common American eel, *Anguilla rostrata*, it could be entering any fresh water from Georgia to the St. Lawrence. The young fish swim so closely that the school seems to be a single endless organism working its way in a generally upstream direction in tidal waters. Eventually some (but not all) of these juveniles tighten ranks and push to the very headwater rivulets of ocean-connected rivers. The fish that got to those headwaters remain there to grow for as many as twenty years and eventually they straggle back downstream as full-size female eels, fat but not developed sexually. Those eels which did not push to the headwaters but remained in the brackish bays and estuaries for several years resting under rocks by day, scavenging by night, growing all the time, turn out to be male American eels. These also become fat, but not sexually mature, and eventually these also straggle toward the saltier waters. All the information currently available suggests that when the grown-up eels enter the sea water they cease to eat and never eat again as long as they live. Those that escape the eel pots in the estuaries and then escape the striped bass in the river mouths and then finally escape the bluefish in the ocean swim back to their birthplace in the mystical Sargasso Sea which really isn't so mystical at all. This 300,000 to 3,000,000 square mile general area of the Atlantic is located approximately between the twenty-fifth and the thirty-first parallels of latitude north and the fortieth and seventieth meridians of longitude west.

This is an area of the Atlantic which gets very little disturbance from winds and currents, but rather, the forces of nature seem to combine to hold captive the very water and its accumulation of plants and slow-swimming resident animals. This area is not especially crowded with life, in fact some authorities consider it to be relatively barren. A desert at sea. Salinity is high in the Sargasso since there is a high evaporation rate and there are no nearby fresh-water rivers dumping into it, or icebergs melting into it. The plant life, and there is really not much of it considering the vastness of the area, is dominated by seaweeds that may have originated elsewhere and drifted in. Suffice it to say here that most of the plants in the Sargasso Sea are found elsewhere as well, but once they arrive, they cease to propagate by roots or shoots or sexual means, but rather as floating bits they just keep growing along. So this is where the grown-up eels go, and having arrived, they mature sexually and then they spawn and finally they disappear. Maybe the old eels go to a graveyard like the fabled elephant graveyard of darkest Africa—well, anyway some way we are rid of them.

Eel eggs hatch in the Sargasso Sea and the larval young look so little like eels that for hundreds of years no one associated these flat, virtually boneless, practically spineless transparent creatures with the slimy eel-like dark round pencil-lead juveniles seen swimming into our river mouths. Zoologists even had another name for these fishlike larvae.

Thanks to money provided by the Danish Beer Trust, brewers of Carlsberg and Tuborg and Elephant label beer, a research vessel was fitted out and maintained to learn where the European eel *Anguilla anguilla* originated. And the scientists did just that. And the hatchery of European eels was found to be in the Sargasso Sea, off Bermuda. The man who spearheaded the study was Johan (sometimes Johannes) Schmidt, and the date was about 1922.

This European eel hatchery area of the Sargasso is approximately but not precisely where the American eel also originates under similar circumstances. Now, the European eel gets to Europe in about three years, working its way north and across the Atlantic generally following the Gulf

Stream. The American eel gets to its "home" waters in less time, perhaps as quickly as six months.

I suggest two interesting questions for your consideration over your next seidel of Carlsberg. First is as to sex in American eels. Could they be hermaphrodites in their youth and could it be that those which remain in the lower estuaries and bays *become* males rather than that they remain because they *are* males? This would be environmental sex determination. There is a carefully reasoned argument for this possibility in the literature. Keep an open mind.

Question number two is whether the American and European eels are truly two species or whether the Europeans simply spent more time as larvae in salt water and thus accumulated more vertebrae? Number of vertebrae is the criterion that separates the species. I might have mentioned sooner in this story that the American eel has 104 to 110 vertebrae and the European sports 111 to 118. Not much else is different. One clue suggesting the hypothesis of one species is that eels from southern U.S. tend to have fewer vertebrae than those captured farther north. Keep an open mind.

If you wish to become embroiled in the eel controversy, you might get started by reading *Nature,* Number 4660 of February 21, 1959 and Number 4695 of October 24, 1959. These two short articles and their references are a good point to depart from.

Fish which breed in fresh water and mature at sea like the shad and salmon are known as anadromous; and those which spawn at sea and mature in fresh water like the eel are described as catadromous. This is good jargon for the trade and should be mastered for your next cocktail party or beer bust.

Let's get back to estuarine and tidemarsh eels. During the daytime, they are usually hidden under debris or stones and at night they are out foraging. As I mentioned before, the resident eels will turn out to be sexually immature males. As they go to sea, they might be hardly larger in diameter than your gloved thumb, but the female might match the diameter of your wrist.

Eels are so important to man in terms of economics, and so dependent on tidemarshes for their development, that I am continuing this discussion to include a few remarks about how they enter our food chain.

Eel meat is readily separated from its bones by the diner, is firm and chewy. At the time of this writing, most eels *consumed* in the U.S. are consumed by blue crabs and striped bass in the form of bait, but in Europe they are a premium-priced delicacy. A vigorous eel export business is developing now and young eels are even being shipped to Taiwan for fattening for eventual sale in Oriental markets. There are only a few minute scales. The skin is usually left on for smoking, but is removed for pickling, stews, frying or other cooking techniques. The viscera is removed and discarded. There is no roe since these fish do not develop sexually until they return to their spawning grounds.

To skin an eel, one must subdue it; no easy feat for the uninitiated. One technique popular in the Connecticut River Estuary which is held in high regard among older rivermen is to place a wad of Mail Pouch or Liberty or Prince Albert chewing tobacco (about a cheekful) in a cup of hot water long enough to let the water cool. The "tea" is then poured into a five-gallon pail half filled with live eels and river water. Soon the eels will be sedated, and it is then that they are beheaded, gutted, and skinned. The job for just a few fish is accomplished with a knife, a pair of pliers and a piece of toweling. You should cut from the throat back through the spine, but do not sever the skin on the back (dorsal) of the neck. Rip the belly with the knife a few inches. Grasp the head with a dry towel and the severed spine with the pliers and pull the parts in opposite directions. The skin will peel inside-out off the meat.

Fins which remain with the meat may be trimmed off with scissors and any viscera left with the torso can be stripped off readily with a push of the thumb. Discard the skin, head and contents of the body cavity. All the meat that remains on the bones is edible. It may be cut into sections or left intact for the cook. Although it is possible to filet an eel, this is rarely done and most cooks and diners consider it unnecessary.

If many eels are to be processed, a machine is desirable. It can be built in a few hours and will last through the lives of several fishermen. The eel, feshly killed or preferably but not necessarily sedated, is placed with its head in the leather-lined cup and a steel plate is lowered to clamp it there. This clamping is equivalent in effect to the grasping of the head by the toweled hand previously described. Then the muscle and bone behind the head is severed with the knife. (This is the cut which certainly kills the eel.) A second cut down the belly a few inches is made to aid in peeling the skin. With the head clamped and the skin at the nape of the neck uncut, the backbone and its meat is grasped by hand, gloved hand, pliers or toweled hand and pulled away from the clamp. Elapsed time per eel is less than 30 seconds using this method. This clamp looks a lot like a mechanical clam shucker but the blade on the eel clamp is 1/8 inch wide and deliberately made dull. It must not cut through the skin at the nape of the neck but it simply holds the head and its attached skin during the process. A recently killed eel which was not sedated will be very active and difficult to hold any other way. The clam shucker blade, by contrast, is very thin and fairly sharp in order to enter an opening between the shells.

FLATFISH

A tide marsh creek with no flatfish is unthinkable. That would be like the kiss without the squeeze or the pie without the cheese. Flatfish are well shaped for shallow water, a winter flounder 12 inches long will weigh over a half-pound, but his vertical height will be only five-eighths of an inch and much of that will be often buried in creek-bottom mud.

There are perhaps ten species found in tidewaters between Maine and Florida. The largest is the fluke or summer flounder, and they have been taken from brackish creeks which measure less in depth than the twenty inch nose to tail length of a two pound fish. Larger fluke than these are found in deep bays and offshore, but there is no question that twenty inch fluke have been taken from twenty inches of water in Great South Bay, Long Island. I know because I caught them fishing with mummychugs between East Rockaway Inlet and Hempstead Bay as a boy. Generally one could see the fish before it took the bait.

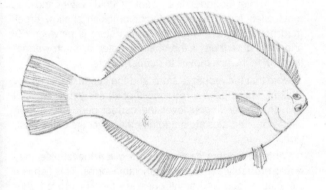

Winter Flounder

WINTER FLOUNDER

The most common flatfish of the U.S. East Coast excluding Florida is probably the winter flounder, also called the blackback flounder. This fish, *Psuedopleuronectes americanus,* is so abundant everywhere and easy to recognize that I will describe all the other species you may encounter by comparison with it. The top color varies as the colors of sand and mud vary and the bottom surface is yellowish or white or blue-white. When you hold the fish with its dorsal fin up and its belly and anal fin down, its dark side is its right side. We call it a right-handed flatfish. In my experience I have found that one in perhaps ten thousand is left-handed, and one in perhaps five thousand has some blotches of dark pigment on its white-belly-blind-side.

The winter flounder likes cool or cold water, and it is most often found in seawater or brackish shallow creek mud bottom waters when the temperature is between 50° and 70° Fahrenheit. If the shallow water is much warmer than 70° F. the fish moves to deeper waters.

Note that the mouth is small and the lateral line is nearly straight. Males have a roughness of scales you can feel if you draw your fingers over the caudal penducle on the blind side. Females are smooth-scaled over the entire blind side.

Winter flounders scavenge, but given the choice they would probably eat nothing but clam worms. This fish is a superlative food and small ones are fried or broiled after gutting (but usually without scaling), and the diner simply peels back the cooked skin and eats the delicately flavored and fine grained white flesh. Much of the fish sold in U.S. restaurants as "sole" is, in fact, filet of winter flounder. Technically, the only American representative of the family of genuine soles *(Soleidae)* north of Florida is the hogchoker, and it has no market value.

SMALL-MOUTH FLOUNDER

This species *Etropus microstomus* is true to its common name, left-handed, large scaled and its lateral line is nearly straight. It is found from New England to Florida, but it never gets larger than about four inches. You cannot confuse this species with the other small-mouthed flatfish, since it is left-handed, whereas the hogchoker and winter flounder are both right-handed.

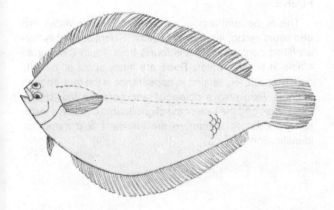

Small-mouth Flounder

WINDOWPANE

This is a left-handed large-mouthed relatively flat flatfish with a high arch in the lateral line as it passes over the pectorals. This fish is also known as the sandflat and sundial. Scientifically it has been referred to as *Lophopsetta maculata* and more recently as *Scophthalmus aquosus*. The windowpane has a relatively large toothy mouth which correctly suggests a diet of fish and shrimp. It is more apt to be found on a sandy bottom than on a mud bottom, but small numbers have been found in tidemarsh creeks from time to time. You will run across specimens from the Carolinas, north. The population center seems to be in Long Island Sound.

FLUKE

This is the common large-mouthed, shallow-water, fish and squid eating, left-handed flatfish. Scientifically it is *Paralichthys dentatus*, and it is found from South Carolina to Maine in shallow water. There are many spots all over its dark side. Closely related in appearance is the four-spotted flounder *Paralichthys oblongus* and here the four prominent spots and the lack of the slight double concaves in the tail of the larger specimens are sufficient field marks for identification.

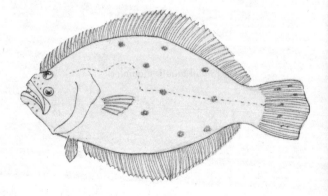

Fluke

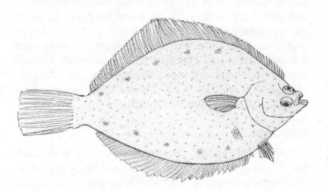

Yellowtail Flounder

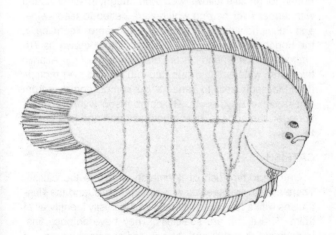

Hogchoker

YELLOWTAIL

You may encounter an occasional juvenile yellowtail flounder, *Limanda ferruginea*, in a tidemarsh creek bottom and you might mistake it for a winter flounder but for its slightly larger mouth and its lateral line which has a high arch over the pectoral fins. It is rare south of Chesapeake Bay but increasingly abundant as you get north. In deep water off Nantucket it is fished commercially. This fish provides most of the "fillet of sole" for the U.S. market and the winter flounder accounts for the rest.

HOGCHOKER

This fish moves from salt to fresh water with no discomfort. It is small mouthed, right-handed and rarely more than about eight inches long. It is transversely striped on its eyed side, and on the blind side it is marked with gray blotchy spots. Its mouth is oddly shaped and a glance at the illustration should suffice. Oral tradition has it that these hard-boned tough little fellows would be caught in seines along with tender river herring which were netted to feed swine, and this is how they got their common name. Technically, the hogchoker was *Achirus fasciatus*, now known as *Trinectes maculatus*, and it is the only true sole of our Atlantic tidemarsh waters. It is distributed in bays and river mouths from Massachusetts to Panama, but is most common from Chesapeake Bay, south. By comparison with the winter flounder, the hogchoker is blunter, striped and less graceful.

CATFISH

One important fish of tidemarsh waters is the catfish. There are several species with similar habits, and the illustrations with a few notes will help you to easily identify all of them. These fish are scaleless. They have adipose fins, barbels at the mouth and their spines can be dangerous if they are handled carelessly. Catfish seem to eat anything they can catch. Shrimp are a favorite food.

The Gaff-topsail catfish, *Bagre marinus* is found from Cape Cod to Panama, but mostly south from Chesapeake Bay. Long streamers on dorsal, pectoral and jaw barbels

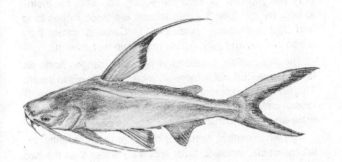

Gaff-topsail Catfish

Sea Catfish

make it unique among our catfish. Their eggs are nearly an inch in diameter and if you have trouble swallowing that, try this: the male incubates a mouthful of these eggs for more than two months, or until the young can shift for themselves. By this time the little fellows are three inches long and Dad is probably exhausted. In Carolina waters they breed in June and July. Adults reach two feet in length.

The sea catfish, *Galeichthys felis* also ranges north as far as Cape Cod but is more abundant from Virginia south. This fish is active at night and is much smaller than the gaff-topsail catfish, and it too is a mouth brooder. Again, it is the male who tends to this chore.

The white catfish, *Ameiurus catus,* or more recently, *Ictalurus catus,* tends to favor less salty water than the two aforementioned species but still it is found in many river mouths and bays from Texas to New York. It spawns in a shallow-water gravel nest and the parents guard the eggs and young. A large white cat might be two feet long and weigh as much as five pounds. The color is blue-gray above and lighter below and it is unspotted. Occasional specimens are slightly mottled but never spotted.

The brown bullhead, *Ictalurus nebulosus,* is generally considered to be a fresh water form, but I have caught them from waters which supported barnacles and even ribbed mussels and blue crabs, so don't be surprised if you meet this fish. The female lays as many as ten thousand small eggs in a nest which both parents guard. The color is olive-green-brown and the tail is square or only slightly notched. This squareness of the tail sets it apart from all the other catfish you are apt to encounter.

Lastly, consider the channel cat, *Ictalurus punctatus.* Its tail is forked even more deeply perhaps than the white catfish which it resembles, but it is always peppered with black or gray specks or spots. This species achieves weights of 55 pounds and the female lays as many as 20,000 eggs in a nest which the male guards. These fish are active predators of all estuarine life, and in turn are eaten by ospreys, otter and of course larger fish and man.

White Catfish

Brown Bullhead

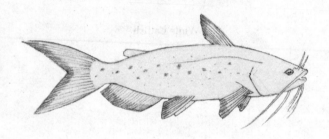

Channel Catfish

CARP AND GOLDFISH

The carp *Cyprinus carpio* is an old-world fish which was introduced to America many years ago. Today it is found in fresh and brackish tidemarsh waters from Georgia to Maine. The same can be said for the goldfish *Carassius auratus*. Their habits are about the same. They hybridize. Perhaps their ancestors were very, very similar.

Today carp have two barbels on each side of the upper jaw and goldfish do not. Also carp have saw-toothed spines at the base of the dorsal and anal fins.

These fish have large scales, small mouths, large appetites—mostly for vegetable matter, and they produce large numbers of eggs—a fifty pound carp may lay seven million eggs in one season. Breeding and feeding and just getting around is a big deal for goldfish and carp. They are frequently seen in marsh creeks tearing up the bottoms, swimming through mud, sometimes with dorsals and tail top showing. I've seen twenty pound carp tearing around in tidemarsh creeks only ten feet wide and ten inches deep. If an osprey ever attacked one of these monsters, I predict he would be taken for the equivalent of a Nantucket sleigh ride.

There are records of carp in Europe spawning in seawater; and in the U.S. every aquarist knows that goldfish do better if the water has a little salt in it; and there is no question that carp are found in blue crab and *Spartina* waters.

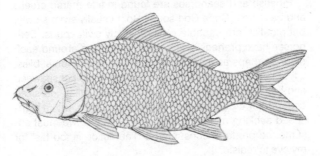

Carp

THE SEAHORSE AND THE PIPEFISH

Yes, these are fishes. No, they surely do not look like fishes, or much act like the others of their ilk. Seahorses and pipefish are not in the same genus, but they are still so similar (and so different from all other fishes) that I lump them here. The seahorse, *Hippocampus hudsonius,* and the pipefish, *Sygnathus fuscus,* are easy to recognize. Unwind a seahorse's tail and he might stretch to seven and a half or eight inches. Stretch a pipefish and it might extend to eleven or even twelve inches. Both are variable in color—chameleon-like, mostly brown, leathery and scaleless. Watch their eyes, both species can roll them independently. The respiration apparatus of both pipefish and seahorse is not quite typically gill-like but rather more tuft-like; however, it serves the same purpose.

Their sex life is not quite fish-like, but it too serves the same purpose. The ripe female deposits her eggs in a slit or belly fold in the male's abdomen and here they develop. They may even get some nourishment from the walls of the pouch which makes it a little like a placental, female mammal except it is still a male fish. Eggs so implanted remain with the male until, as tiny replicas of the parents, they hatch and swim; and sometimes for a while, according to some reports about the seahorse, they come and go as they grow. This is somewhat like a kangaroo except that it is the *male* who provides the pouch.

Pipefish and seahorses are found in tide-marsh creeks and bays from Cape Cod south. Both usually swim slowly, but pipefish are capable of surprisingly swift spurts. Seahorses have prehensile tails which they wrap around each other, perhaps as a gesture of affection or connubial bliss. Other times they use their tails to attach to a bit of seaweed and then they remain virtually motionless until a flea-sized copepod or crab or shrimp comes close; then, with a high-speed sucking motion, the prey is gone and we must assume the predator consumed it. The action is too fast for my eye to register it.

Pipefish and seahorses do well in aquaria, and males with full belly pouches will "incubate" or brood their progeny in captivity. Brine shrimp eggs *(Artemia salina)* are available in pet shops and are easy to hatch into excellent food

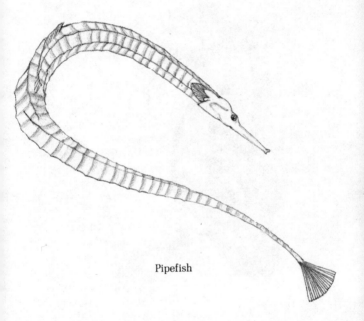

Pipefish

for the young. The newborn fish will probably starve unless provided with small shrimp, small copepods or other small-size food, but the adults can manage young guppies and other organisms in that size range.

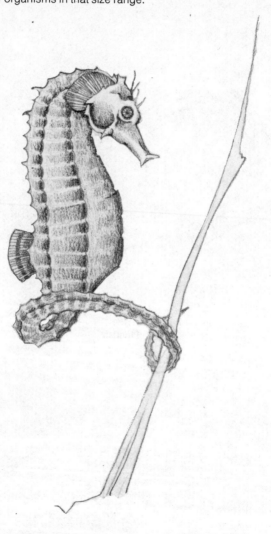

Seahorse

STRIPED BASS

Also known as rockfish, linesides, rock and striper, *Roccus saxatilis,* is probably the best known and most avidly sought game fish produced in the tidemarsh waters. It is not hard to recognize. The only fish very much like it is the white perch, and even here the distinction is clear. The two dorsals of the striper don't touch but on the white perch, they do touch. Additionally, the largest white perch hardly exceeds two pounds, but the record striper according to Smith in the North Carolina Geological and Economic Survey, Volume 2, 1907, page 271, is about 125 pounds. Another, weighing 112 pounds, was taken at Orleans, Massachusetts, according to Bigelow and Schroeder, *"Fishes of the Gulf of Maine,"* U. S. Fishery Bulletin No. 74, 1953.

Females tend to be larger than males, but regardless of sex, they are so powerful that they are often found feeding actively in a roaring surf. Many a fisherman has seen a striper in the crest of a comber a moment before it broke. Stripers breed in brackish or nearly fresh tidal creeks and rivers. Most spawning takes place between New Jersey and Alabama, with the Delaware and Chesapeake tributaries probably the greatest producers. There is another population base in the Gulf of Mexico and still another (from introduced fish) in the Pacific between Los Angeles, California and Greys Harbor, Washington.

Stripers live a long time. One in captivity lived 23 years. And, in 23 years, not counting the first four growing up, stripers produce great numbers of eggs. A four and a half pound female was found to be carrying 265,000 ova and many larger specimens may well spawn over a million eggs per year. The eggs are semi-buoyant, and they gradually drift downstream from freshwater to tidewater. All evidence suggests that, depending on temperature, they hatch in about three days and the larvae and juveniles fend for themselves. Stripers have been studied intensively and fished intensively, but there is still much about their routine habits we don't know. Perhaps our lack of understanding stems from our assumption that they are creatures of habit.

I suspect they have few hard and fast habits but rather they are versatile opportunists. Some stripers are found land-locked in fresh water lakes. Others are caught in the surf while feeding on crustacea and mollusks and still others in tidal creeks where they come to eat sand eels, *(Ammodytes),* eels, grass shrimp, crabs, worms, and silversides.

The only pattern we are really sure of with stripers is that they do not spawn in salt water.

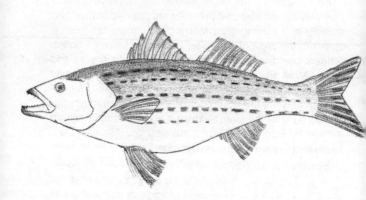

Roccus, Striped Bass

Note the space between the dorsals.

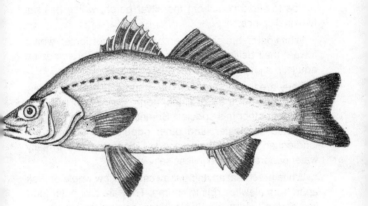

Morone, White Perch

Note that the dorsals
touch each other.

WHITE PERCH

This is a brackish water fish and a dominant predator in tide-marsh waters. First glance suggests a small version of the striped bass. The largest white perch (*Morone americana*) might weigh two pounds, and measure fifteen inches in length.

To identify this fish in tidemarsh waters, look for the two touching dorsals. The first one has nine spines and rays. This is all you need to notice to separate it from the striper (*Roccus saxatilis*). Then look at its anal fin. There are three spines here, in front of the soft rays. That is all you need to notice to separate it from the silver perch, which has but two anal spines.

White perch are gray-green on top and get lighter, whiter toward their bellies. There are no spots, bars or stripes on adults, but juveniles do have pale longitudinal stripes.

They are found in creeks, bays and estuaries, generally in schools from the Carolinas, north. Salinity, or lack of it does not bother this species. Sometimes they are found in undiluted sea water, and other populations of them are landlocked in fresh water ponds. Those that live in salty water generally spawn in less salty water.

White perch eat anything they can swallow whole or pick apart from alewife eggs to shrimp. They are a popular food fish and will strike at artificial lures, so anglers are always looking for them. Chesapeake Bay produces over a million pounds per year for human consumption.

STICKLEBACK

There are at least four species of sticklebacks to be found in our East Coast saline, brackish and freshwater tidal marsh creeks and bays. The largest specimen of the largest species might reach four inches long and the smallest half that. They vary in color depending on sex and season but are mostly brassy and mottled gray with variable rosy markings on nuptial males. These little fish are pugnacious and eat anything animal or plant that they can swallow or pick apart. Most, but not all species choose relatively fresh or slightly brackish water for breeding; the remainder of the year they are more likely to be found in saltier waters.

The genus takes its common name from spines which may be raised and locked in place. Two spines swing out from the sides of the abdomen as substitutes for ventral fins. In some species there is one additional spine in front of the dorsal fin. They provide this little fish a formidable defense. The illustration will help with identification, but the number of spines you count is no sure thing. That should be no surprise; most things in nature are not sure things. Have you ever heard of a fish that builds a tunnel of love?

The male sometimes builds a nest. It is made of bits of water plants and their roots. He entices the female to enter. She spawns. He fertilizes the eggs and drives her away (lest she eat the eggs), and he guards the nest until the eggs hatch and the young swim free. This might take all of twelve days. During this time it is possible that the female might take her additional wares elsewhere.

Let's look at that tunnel nest again. It is on or near the bottom, barrel shaped, about an inch in diameter and the bits of vegetation from which he made it are cemented together. The cement is a mucous which he produces in his kidneys and which he secretes from a pore near his vent as a thread, to bind the nest together. He weights the nest down with grains of sand if it appears too buoyant and then—no, I didn't dream this up, Mother Nature did—he entices as many females as he can into his architectural masterpiece and there each female lays as many as 150 eggs which he promptly fertilizes. The exact number of females he entices is his business and a strenuous business it is. When the eggs are about hatched, he destroys the nest but guards the larval young until they shift for themselves. By this time, many males are so worn out they die. Those that live, go back to saltier waters until the next time they get that urge.

In Alaskan waters one study suggests that sticklebacks are hermaphrodites and fertilize themselves. Or maybe they are a race of females or perhaps the females are viviparous, or simply ovoviviparous. In any event, some fish had to have been fertilized internally (like guppies) since developing embryos were found in the abdomens of female

Alaskan sticklebacks.

The family of sticklebacks is known as *Gasterosteidae* and the genera you may encounter are *Apeltes* and *Gasterosteus*.

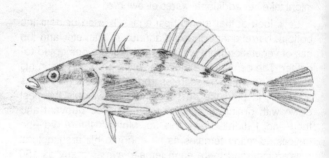

Stickleback

SAIL-FIN MOLLY OR SAIL-FINNED KILLIFISH

This is a three or possibly four inch, common, live-bearing, brackish water, mosquito and algae eating, denizen of fresh and tidemarsh waters, from South Carolina south. This fish is less avid a feeder on mosquito larvae than is the mosquito fish, *Gambusia.* In fact it will survive for a long time on algae alone. The dorsal fin on the male is its claim to fame and like the lower jaw of the anchovy you must see it to believe it. Aquarists have domesticated this species and have established several strains with colors not seen in nature, but regardless, the male alone sports the tremendous dorsal fin. Scientifically, this species is known as *Poecilia (Mollienesia) latipinna.*

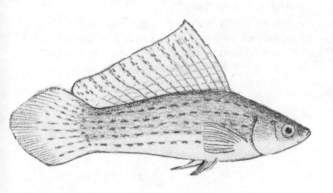

Sail-fin Molly

MOSQUITO FISH-TOP MINNOW

The males are but an inch long and the females twice that. They are shaped like guppies, they give birth to living young. Baby mosquito fish are born a few at a time, each perhaps a third of an inch long.

Mosquito fish *Gambusia affinis holbrooki* are prolific and hardy and tolerant of brackish to fresh waters. They are rare in New Jersey, straggling in Delaware and abundant from Virginia south. They are sometimes released in waters north of New Jersey in the spring and of course they survive through the entire mosquito season and sometimes even perhaps a mild winter. This fish is aptly named. Its major item of diet is mosquito larvae and it eats tremendous numbers. Surprising for so small a fish. Its size is actually an advantage since it can get into grassy areas not accessible to larger fish. Aquarists and mosquito control commissions frequently release these fish outside their natural range, so don't be surprised to find them where you might be surprised to find them.

Another similar livebearer is the dwarf top minnow or mosquito fish, *Heterandria formosa*. It is smaller than *Gambusia* and is found from North Carolina south through Florida. It is less tolerant of salt water than is the *Gambusia*.

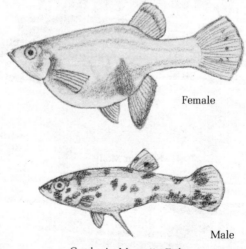

Female

Male

Gambusia, Mosquito Fish

SNAPPER BLUEFISH

The snapper I'm writing about is the juvenile bluefish *Pomatomus salatrix.* Bluefish grow to nearly thirty pounds and are found from Maine to Florida. The first dorsal is spiny, the second dorsal is soft. There are no tiny finlike rays behind the dorsal called finlets (the mackerel and the pompanos have finlets). The tail is forked, there are no spines associated with the anal fins, and the scales are relatively large. The color is dark gray-blue above the silvery blue below, and there are no spots or stripes or blotches.

The adult bluefish is an occasional visitor to tidal creeks—in a frenzy he will drive mullet, silversides, sandeels, alewives, blueback herring or menhaden anywhere, even into fresh water or onto a shoal. It is considered by many authorities to be the most bloodthirsty fish in the sea. And what bluefish don't eat, they mangle, or cut up with "U" or "V" shaped notches. They will attack anything that moves.

Bluefish favor water warmer than 58° F. and the adults are found on the oceans, by the coasts and in estuaries. It is the younger bluefish that is usually encountered by the visitor to the tidemarsh. His food supply is abundant in tidal creeks, and young blues, hatched at sea, congregate in tight schools frequently in brackish tidewaters, and occasionally even in fresh water. Small snapper-blues feed on copepods and larvae of mollusks and crustacea as well as on typical baitfish such as anchovies, silversides, and juvenile herrings of all species.

Young snappers grow rapidly in the tidewaters and in the course of a summer between Cape Cod and Hatteras many a girl or boy has seen them grow while school was out between Independence Day and Labor Day. Snapperblues are fun to catch with any moving bait or shiny lure and are delicious fried, even when only six inches long.

When bluefish get into a school of bait the water seems to boil. Gulls dive into the water for scraps and terns work the edges for small fish which were chased to the surface. In Florida, (and I know the Chambers of Commerce will hate me for mentioning this), bathers have been bitten and severely injured by adult bluefish in only three feet of water.

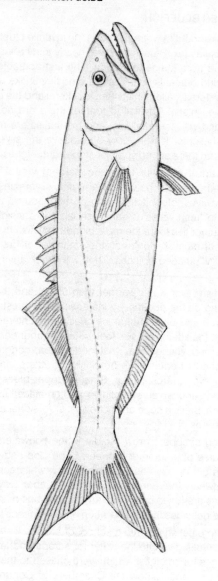

Bluefish

ANCHOVY

A silvery schooling tidewater fish up to six inches long with large eyes, wide silver stripes from gill cover to base of tail, and only one dorsal fin, this is the anchovy. The single dorsal and the width of the silver band set it apart from silversides, the only other tidewater fish it even remotely resembles. Another distinguishing feature of this genus *Anchoa* is the tremendous lower jaw. You have to see it to believe it, and once you see it you will not forget it.

Anchovies are of several similar species and are found along our entire seaboard, but they are only occasional north of Chesapeake Bay; to the south they are abundant.

Their habits are much like "silversides" (*Menidia* species) but they tend to stay clear of the *Spartina* and so are of less interest to a tidemarsh visitor.

One regrettable thing about the anchovy is its common name, whitebait, which is also a common name of silversides. The flesh of the anchovy is much more oily than that of the silversides.

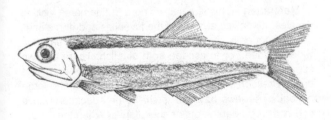

Anchovy

HERRING

Herrings are blue and silver-blue fish with large loose scales, a single dorsal fin, no adipose fin, a forked tail and few or no teeth in their jaws. They swim in large schools with their mouths open. Some are anadromous, that is, mature fish lay their eggs in fresh water—streams and rivers, then the juveniles make their way back to sea where they mature. Some adults of some species survive the effort of spawning and these return to sea and come back to spawn again and again while others die after completing one cycle. The tidemarsh creeks see some of them three times, once as adults on the way to spawn and again in midsummer as some survivors go back to sea and in the autumn as the fingerling juveniles slowly work their way into saltier waters to grow up. Let's look at the various species of herrings in terms of the tidemarsh.

Atlantic Herring - A purely pelagic (sea-going) form, *Clupea harengus,* does not normally enter non-saline waters. This is the true herring of commerce, the one that helped make Norway famous. Relatively long, slender, roundish. It has teeth you can feel in the roof of its mouth. Its tongue shows when it is viewed from the side with its mouth open. It is a rare visitor to brackish waters, and would be a surprise in a seine drawn through a tidemarsh creek although I have found them in brackish shallow bays.

Menhaden - This is a sea-going and tidemarsh visiting species. It spawns at sea but the juveniles and sometimes adults frequently move into low salinity or even salt-free waters, to feed or to escape from bluefish. There are two species of menhaden on our Atlantic Coast. The dominant one is *Brevoortia tyrannus* and it is deep bodied and large headed. The jaws have no teeth. Some specimens have one or several dark spots on their flanks. When the mouth is opened, the tongue shows in side view. Schools of tightly packed fish cover acres and sometimes literally fill river mouths and bays. If they were driven there and bottled up by bluefish, chaos results. No sense in trying to describe it, no one believes it 'till he sees it for himself. Menhaden weigh hardly more than one pound each but by numbers or by total weight, they are the dominant fish of the seas.

Atlantic Herring

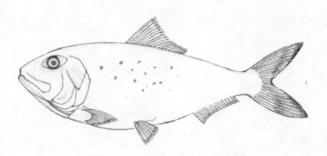

Menhaden

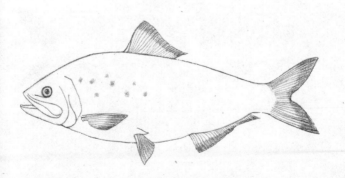

Shad

Alewife

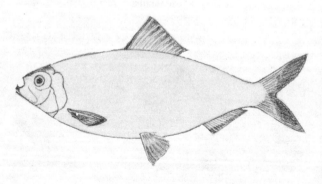

Blueback Herring

No other single species of fish has as many individuals or as many pounds alive.

All herring are oily and menhaden are especially oily. A school actually leaves a slick on the surface of the water which experienced fishermen watch for and recognize. Menhaden have thirty common names between Maine and Florida. They are frequently called bunkers, moss bunkers, fatbacks, or pogy (not porgy). Large schools are watched by fishermen who know that they will be followed by bluefish, weakfish or striped bass. Small menhaden are preyed upon by white perch and squid. Adults at sea are eaten by tuna and porpoises.

When bluefish work on menhaden from below, the birds move in to pick up scraps from above. A regular nautical brouhaha results with bloody water, screaming birds and splashing and jumping fish.

The menhaden is easy to net since it does not dive deep, nor do schools break up under stress, but rather they close ranks and frequently remain on the surface. This makes them easy to capture and form the basis of a considerable industry.

Menhaden are cooked industrially to extract fish oil which is then used in paints, and soaps. Glycerine is another product worth noting. In old fashioned wars, the glycerine which is produced when the fish oil is reacted with alkali was invaluable. It is combined with nitric acid to create the famous, however unstable, nitroglycerine, and other more stable explosives as well. A substantial part of the destruction during the First and Second World Wars was derived from this oily toothless fish. The solid which remains, after the oil is removed, becomes the fish meal which is used in cattle feeds.

The menhaden consumes vast quantities of phytoplankton which was produced on the tidemarsh. This fish extracts his food from the water by swimming rapidly with his mouth wide open, and the tiny organisms pile up on comblike gill rakers arranged on the arches which also support his gills. There are no teeth on his jaws.

From time to time, generally in late summer, the vast shoals of menhaden in tidewater bays and estuaries sicken and die. The birds cannot keep up with the food supply and soon the newspaper feature writers close in to pick off a few choice attacks on some polluter who didn't necessarily contribute anything to the demise of the fish. When a school of twelve-inch fish, packed so close they seem to touch, covers as much as ten acres it is spectacular in life and in death. Government data for 1946 lists the menhaden catch for the Atlantic and Gulf coasts for 851,129,000 pounds (over 425 thousand tons) in one year!

The American shad, *Alosa sapidissima,* famous for its roe and boned meat grows to at least ten pounds and enters rivers from Massachusetts to Virginia to spawn. The shad "run" of ripe adults upstream takes place in springtime and the date is determined by the temperature of the water. 50 to 55°F seems to trigger their trip. In the northern rivers up to 20% of the adults survive to return to sea and try again the following year. Warmer southern streams have fewer returnees.

Young shad move downstream in late autumn first eating plankton and then larger organisms as the grow. They remain at sea for several years before attaining adult size and that inexorable urge to do their thing. The shad has a few teeth in its jaws which you might be able to feel, and a few more in the roof of its mouth. Its tongue shows when viewed from the side and its flanks are generallly spotted. A similar species is the **hickory shad,** *Alosa mediocris.*

Shad are netted and hooked only during their spawning run upstream. A "spent" shad is a miserable excuse for a fish and only the birds bother with these weakened and wasted stragglers.

Boned fillet of shad is a great delicacy, and the art of preparing it is acknowledged a great art indeed and from the Susquehanna to the Connecticut there are people who have mastered the art of deboning this delicious but very bony fish. The unripened eggs (roe) of the shad also makes a great dish, generally broiled.

Two more herrings which enter the tidewater are the **blueback** and the **alewife.** Let's consider them together since most laymen and even many baymen never bother to separate them. The blueback herring, *Alosa aestivalis,* has a blue-black lining to its abdominal cavity and the alewife, *(Pomolobus or) Alosa pseudoharengus* is pink or pink-palegray in the same place (the peritoneum). Both grow to eleven or twelve inches and weigh not much more than a half pound as adults. The tips of the tongues of both species are hidden by the jaws when their mouths are opened and the alewife has larger eyes than does the blueback. In fact, the alewife is called the buckeye (buck eye) because of its large eyes. Also, alewives are sometimes found in fresh water lakes as "landlocked" and they grow to maturity and even reproduce without any salt water in their life cycle. Under ideal anadromous conditions, both species survive the spawning and return to sea in early summer.

Alewives and bluebacks are also plankton eaters with highly developed gill rakers and no teeth in their jaws. The identification of these river herrings is easy for anyone who has worked with them for a few years. Others should be cautious or play it safe by simply saying "river herring" or "anadromous herring" since this appellation covers all but the easily recognized Atlantic herring and the menhaden.

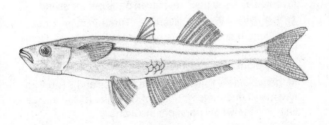

Silverside, *Menidia*

SILVERSIDES

There are several species of tidemarsh silversides all so similar that only ichthyologists try to separate them. Even the specialists have trouble (or fun) since these species tend to grade into each other. Suffice it here to say that tidemarsh silversides are found in *Spartina* marsh waters from Florida to Maine. They congregate in same-size schools and they tend to act in unison. This is in contrast to the disorganized mob scene of a *Fundulus* school. The narrow silver stripe from the high pectoral fin to the base of the tail is the first field mark. This silver stripe is outlined on top by a narrow black streak. Note also that all silversides have two dorsals. The other tidemarsh silversided school fish you might confuse it with is the anchovy, but anchovies have just a single dorsal fin, the pectoral fin originates low on the belly, and the silver stripe covers about 25% of the side of the fish.

Silversides are all in the genus *Menidia* and each of the three tidewater forms can tolerate the full range of salinity from fresh to seawater.

M. menidia - Green smelt, sand smelt, white-bait, capelin, sperling, spearing, shiner. This species reaches a length of five and a half inches and has 23 to 26 anal fin rays. It is common from Newfoundland to Florida and its favorite salinity is near seventeen parts per thousand.

M. beryllina - Waxen silverside. Three inches is maximum length and anal rays number only 15 or 16. Common from Massachusetts to Mexico, its favorite salinity is near eight parts per thousand.

M. peninsulae - This is a Florida species and it is unlikely you will encounter it north of Georgia. Its favorite salinity is near twenty parts per thousand.

Silversides are not especially important as a cash value product, but they are very important in the food chain. They eat all small things including their own eggs, shrimp, copepods, amphipods, young squid, worms, insects, algae and diatoms. They are eaten by man — usually fried as "whitebait," and by virtually every predatory fish that frequents or visits the coastline. Bluefish, mackerel, striped bass and white perch eat tremendous quantities, and so this genus occupies a vital pivotal role since it consumes some single cell plants and animals and many small and simple forms, and it in turn is eaten by creatures like man and other ultimate consumers which in turn are not eaten but simply die and decompose to feed scavengers and bacteria. This comment is applicable also to the anchovy.

YELLOW PERCH

This is a common fresh water fish usually less than a foot in length. It has two dorsals, a slightly forked tail and it is usually a shade of yellow or yellow-green with about seven broad but not necessarily sharply defined vertical bars spaced along each side. The first dorsal is spiny and after a short gap, there is a second dorsal of one spiny ray and about 14 soft rays. This is *Perca flavescens* and I mention it only in passing because it is rare in waters where the saltiness can be tasted. It will be found from the wild rice upstream, overlapping the white perch, but only now-and-then in creeks where species of *Spartina* grow on the shore.

Yellow perch range from Nova Scotia to North Carolina and are often preyed upon by ten-year-old boys who are just beginning their angling experience.

JOHNNY DARTER

Three inches, fast, inconspicuous, this is a solitary fish of nondescript mottled browns. It has two dorsals, one anal fin, and a rounded tail. Call it *Boleosoma nigrum*. There are many similiar species of the genus *Boleosoma* with widespread distribution and like the related yellow perch, they will sometimes show up in the slightly brackish water of a streamfed tidemarsh.

STURGEON

This is a primitive, long-snouted fish furnished with barbels in front of its protrusible bottom-feeding mouth. It is armored with bony scales. Maximum recorded size is eighteen feet, but today eight feet would be a whopper and even it would weigh several hundred pounds.

You may find a few juveniles when you seine anywhere in tidemarsh waters between the St. Lawrence and the Carolinas, but generally when these fish are not at sea, they tend to rivers and bay bottoms rather than tidemarsh creeks.

Scientifically, the common sturgeon is *Acipenser oxyrhynchus,* and it is the adult female who produces the famous roe called caviar. This sea-going species is, like the striped bass, anadromous; eggs are laid in fresh water. Young sturgeon spend their first year or more in the lower reaches of rivers before going to sea for the remainder of their lives. Adults return to sea promptly after spawning, and seemingly they repeat the process annually.

With respect to the famous old barroom ballad about "the virgin sturgeon that needs no urgin'," I should mention that a female sturgeon never needs no urgin', but acts about the same way during *each* and *every* spawning campaign, even when she is fifty years old.

AMPHIBIANS

Toads, frogs, salamanders, sirens and other cold-blooded, legged, tailed or tailless, scaleless, warty or slimy or non-slimy creatures of wet or damp places are not found in seawater-salty waters. However, in 1958 W.T. Neill wrote a paper entitled "The occurence of amphibians and reptiles in saltwater areas..." Mar. Sci. Gulf Carib. Bull. 8 (i): 1-97 and in it he stated that "the following species are known to occur in direct association with brackish waters". Note, Neill says *brackish.* How salty is brackish? Well, it varies. Here is Neill's list with my brief descriptions.

Ambystoma maculatum, **spotted salamander,** Nova Scotia to Georgia, two rows of yellow or orange spots on a black back, six to eight inches.

Siren lacertina, **greater siren,** Washington, D.C. to Florida. External gills which cover tiny forelegs. Nondescript olive variable coloration. Twenty to thirty inches and no rear legs.

Acris gryllus, **southern cricket frog,** from Virginia, south. Small (to 1¼ inches) warty non-climber with very variable coloration.

Bufo quercicus, **oak toad,** from North Carolina, south. Small (to 1½ inches) multicolored, warty, non-climbing.

Gastrophryne carolinensis, **eastern narrow-mouthed toad.** From southern Maryland, south. Small (to 1½ inches). Color variable. Habits: retiring, favors shelter and moisture.

Hyla cinerea, **green tree frog.** Delaware to Florida. Usually green sometimes yellow or grey-green. Several subspecies. Up to 2¼ inches.

Hyla fermoralis, **pine woods tree frog.** Maryland to Florida. To 1½ inches. Difficult to identify without a specialized field guide. Color varies from gray to reddish brown.

Hyla squirella, **squirrel treefrog.** From Virginia, south. To 1½ inches. Wide range of colors. Identification in field is difficult.

Limnaoedus ocularis, **little grass frog,** also sometimes called *Hyla o.* from Virginia south. To 5/8 inch. A dark line through the eye on a really tiny tree fog.

Rana clamitans, **green frog.** From North Carolina, north. To 3½ inches, this species looks like a small bullfrog but it has two folds of skin on the body, running back from the eyes.

Rana catesbeiana, **bullfrog.** From Nova Scotia, south. To six inches. Green or grey-green. No skin folds on trunk of body by comparison with *R. clamitans*. Edible.

Rana pipens sphenocephala. This is the **southern leopard frog** and the sub-species may not be valid. From New Jersey, south. To 3½ inches. Color variable but with a few large dark rounded spots having light colored borders.

Also, according to J.D. Hardy, Jr. Chesapeake Science Volume 13, the following "have been collected from barrier beach ponds which are at least occasionally subject to salt spray..."

Ambystoma opacum, **marbled salamander.** From New England to Florida. To 4½ inches. Silvery. Sometimes striped or spotted. Female guards eggs. Similar to *A maculatum* except marks white rather than yellow.

Bufo woodhousei fowleri, **Fowlers toad.** Resembles the very common American toad except that many have a single dark breast spot. From Virginia to New Hampshire. Each dark spot on its back may have three or more warts.

Acris crepitans, **northern cricket frog.** From Virginia to Long Island. To 1 3/8 inches. Subtle differences in markings on thigh required for identification.

Hyla crucifer, **spring peeper,** a treefrog. From northern Florida, north. To 1¼ inches. Brown or tan with dark X on back.

REPTILES

Cold-blooded, lung-breathing, scaly; the reptiles are represented on our tidemarshes by turtles, lizards and snakes. The only two North American crocodilians are the crocodile and the alligator, but alligators, which may range as far north as the southern border of Virginia, prefer fresh water and the crocodiles are limited to south Florida saline and brackish water where they are rare, endangered and nearly extinct.

Reptiles have little or no control over their body temperature and since (except for the sea turtles) they don't migrate, those that live in places that get cold must mark time until the temperature rises. Call it hibernation.

THE COMMON SNAPPING TURTLE AND THE
UNCOMMON DIAMONDBACK TERRAPIN

Is there something nice to say about the common snapping turtle? Not much. It is a scavenger and is sold in some parts of this country for food; but on the East Coast of the U.S. in 1977, when the common snapping turtle *Chelydra serpentina* is weighed in the scales of human public opinion, it comes up wanting.

This is one of our largest regularly seen tidewater animals. Brackish tidewater creeks only twenty feet wide and varying in depth from two to ten feet will support thirty-pound specimens. A meandering creek two miles long might easily support a dozen large adults and a hundred juveniles. The common snapper is found in brackish and fresh slow moving or still water from north Florida to Canada. A subspecies is found in the remainder of Florida; and the Gulf Coast is populated by the "alligator snapping turtle," a member of another genus. The alligator snapping turtle is perhaps the world's largest fresh water turtle and is reported to reach 200 pounds but is less prone to inhabit tidal waters than is the common snapper, so we need not consider it any further.

The snapper will eat anything he thinks he can digest and he is an incurable optimist. He is not a dainty diner. He is known to consume vegetables and live ducklings. Also in

Common Snapping Turtle

Diamondback Terrapin

his diet are carrion, fish, crabs, mussels, fishing tackle and smaller turtles.

This species is an article of commerce, and tidewatermen do ship a box of snapping turtles to the New York or Baltimore food markets from time to time. Generally these are harvested from the bottom of soft mud creeks in cold weather. The method is to probe for them with a pointed stick while they are dormant or hibernating. In summertime and in warmer latitudes throughout the year they are trapped or taken on trotlines. The liver is oily and the meat is flavorsome but much of it is tough and stringy. Pressure cooking helps. The thoroughly cooked snapper produces a large amount of jelly which thickens soups and is acknowledged to have a desirable flavor.

The female lays her eggs, as many as forty, nearly white, up to 1¾ inches in diameter in a nest above the highwater mark. The eggs hatch after about 80 days or the following spring if they were laid in late summer. Except for egglaying, the snapper is aquatic. It eats underwater and it frequently fights and mates in the water.

Snapping turtle are sometimes seen in fighting or mating postures in shallow tidal rivers. When two such biting, clawing, thrashing, hissing, splashing armored tanks confront each other in a twenty foot wide creek one would think the Martians had landed. The riotous uproar is beyond my ability to describe or yours to believe until you see and hear it for yourself.

The enemies of the snapper are raccoons and skunks which eat the eggs and hatchlings. An adult has no one but man to fear. If not run over by an automobile or caught by a game manager or waterman, he might live fifty years. The average common snapping turtle is riddled with parasites. Leeches attach to their skin, and heartworms and nematodes are often found inside.

DIAMONDBACK

Tidal marshes support another turtle genus so outstanding it deserves our rapt attention and silent admiration. Anything more disturbing than raptness and silence will send it scurrying away underwater for parts unknown.

There is but a single species in the genus found from Cape Cod to the Florida Keys, and, within it, seven recognized races. *Malaclemys terrapin terrapin* is the scientific name and the common names and races include northern diamondback, Carolina diamondback, Florida East Coast terrapin and mangrove terrapin. Still other races are classified within a different subspecies and these are found in Gulf Coast waters.

The top shell (carapace) of the largest female is about eight and three quarters inches long and largest male might make all of five and a half inches. Not much to eat, considering all the attention the diamondback has been given. The color is extremely variable but the throat is white or creamy with black spots or blotches. The jaws are frequently white and the shell is attractively sculptured.

Since the common snapping turtle and relatively uncommon diamondback terrapin are the turtles most commonly found in Atlantic Coast brackish tidemarsh waters, it might be convenient to tabulate and compare the salient features of two specimens of the same size.

Snapper	Diamond Back
Color Mud-Brownish-gray	Gray-green. Prominent lips. gray (male) or white (female) and white or cream throat with black markings.
Skin Spiny, horny, warty	Fine-grained, even texture, leathery
Tail Prominent, saw-toothed, long	Relatively small, smooth
Undershell (plastron) Tiny, cut out for legs	Large, notched only for tail
Top shell (carapace) Saw-tooth in back	No saw-teeth
Overall Appearance A really ugly turtle	A really pretty turtle

Malaclemys is *the* aquatic brackish water terrapin. A box tortoise or musk or mud turtle may appear on a marsh border from time to time, but the diamondback *chooses* a brackish tidal creek with peaty banks and a mud bottom and there is where he stays. Diamondbacks eat all manner of animal and plant life, but their natural food is crustaceans and mollusks with a smattering of insects taken at high tide and perhaps a scavenged fish from time to time. They eat under water, and although they will survive in fresh water, they do not choose any but the brackish. The U.S. Bureau of Fisheries established an experimental station at Beaufort, North Carolina and mastered the art of rearing these handsome and good tasting turtles for human consumption. They created a "hybrid race" by crossing northern turtles with a larger but otherwise less desirable southern race. The study continued for nearly forty years and much was learned, but to this day there is no booming diamondback terrapin industry. Lillian Russell and Diamond Jim Brady are both dead, the "Gay Nineties" are over, and what with taxes, there isn't enough money left over to spend it on the meat of an eight inch turtle. When one pound (including skin and much bone) diamondbacks cost $90 per dozen, alive, wholesale; and good cuts of boneless prime beef were only 20¢ per pound at retail, there was a justifiable snob appeal in terrapin as a specialty food. Today when you can spend several dollars per pound for not-too-special steak you don't need terrapin to be a successful snob. As a matter of fact there are very few people today who can consistently recognize the difference between the taste of the diamondback and that of some of the other larger and more common edible turtles.

This turtle hibernates in the northern part of his range and is inactive everywhere during cold spells. When he is cold he can be found and captured from mud in creek bottoms. Eggs are laid in nests excavated by the females on high marshes. Up to a dozen one-inch nearly-spherical white eggs are laid and they hatch in about three months, or the following spring if winter inhibits the completion of their development. A female starts laying in her fourth or fifth year, hits her stride at age 25 and continues to lay until she is perhaps 40.

Diamondback terrapins are preyed upon by rats, muskrats and raccoons who especially relish their eggs. Crows, herons and snakes will eat the young if the occasion permits, but the real limits on their population are habitat and man.

MORE REPTILES

I have already mentioned the two reptiles which make brackish tidemarsh creeks their year-round home. Of those two, the diamondback terrapin is the most closely tied to the brackish tidewater marsh since he is not found in fresh water. The snapper, of course, is ubiquitous. Here, let us take a quick review of the other reptiles you may by slim chance encounter.

SEA TURTLES

Adult female sea turtles visit beaches and tidewaters to bury their eggs. You may also see the newborn young as they hatch and scramble back into the water.

When I was a boy i fished for weakfish *(C. regalis)* off the south shore of Long Island and had the misfortune to lose a year's growth when a sea turtle half the size of my rowboat silently surfaced alongside and took a long deep hissing breath. Totally unexpected, the experience was unnerving to say the least. We never saw each other again.

Caretta caretta: **atlantic loggerhead** Its reddish-brown color is unusual for sea turtles. To 900 pounds. Found north to Newfoundland but nests from North Carolina, south. There are but two claws on each limb.

Chelonia mydas: **atlantic green turtle** Brown (the fat is greenish). To 850 pounds, the usual weight ranges between 120 and 200. Stragglers from Massachusetts, south. Usually there is but one claw per limb.

Eretmochelys imbricata: **atlantic hawksbill** This is the tortoiseshell tortoise. Up to 100 pounds. Stragglers from Massachusetts, south. There are usually two claws on each limb.

Lepidochelys kempi: **atlantic ridley** Looks like a small cross between a loggerhead and a green turtle, but it isn't. From Nova Scotia, south. There are three claws on each limb.

Dermochelys coriacea: **atlantic leatherback** May weigh up to a ton! This is the largest living turtle in the world. There are seven prominent ridges on the upper shell (carapace). Sometimes nests in Florida but is found swimming as far north as Nova Scotia.

TURTLES WITH HINGED PLASTRONS

Kinosternon subrubrum - **eastern mud turtle.** To four inches. From Connecticut, south. High smooth carapace.*Two* hinges on plastron. Dark and nondescript.

Sternothaerus odoratus - **stinkpot.** To five inches. From Florida to Maine. A small smelly musk turtle with light colored stripes on its head and barbels on its chin and on its throat. The plastron has but *one* hinge. The carapace shell is dark smooth and highly domed.

Terrapene carolina - **box turtle.** To six inches. From Georgia, north. A single plastron hinge. Slightly sculptured carapace usually decorated with yellow or orange pattern markings. A common land turtle, it is sometimes found soaking itself in brackish water mud.

HARDSHELL TURTLES WITH NO HINGE ON PLASTRON

Chrysemys picta - **eastern painted turtle.** To six inches. From North Carolina to southern Maine. Smooth shell. Dark carapace with red marks on margin. Light colored plastron. Several subspecies.

Chrysemys rubriventris, Pseudemys r. - **red-bellied turtle.** To fifteen inches. From North Carolina to New Jersey and an isolated population in Plymouth County, Massachusetts. Like the painted turtle but plastron has a red pattern on an amber background.

Chrysemys scripta, Pseudemys s. - **red-eared turtle.** To eleven inches. This is the famous pet shop turtle and therefore escapees and local populations may show up anywhere. The plastron is usually a patterned green and the red or orange ear patch is a good field indentification mark.

Deirochelys reticularia - **chicken turtle.** To ten inches but usually six or less. From North Carolina, south. Striped on neck and on hindlegs. Esteemed as food.

Clemmys guttata - **spotted turtle.** To five inches. From Georgia, north. Yellow polka dots on a smooth dark carapace.

LIZARDS

Some are legless but all have movable eyelids and external ear openings.

Ophisaurus attenuatus - **slender glass lizard.** To eleven inches. From Virginia, south. This lizard is legless and its long tail is brittle. There is a deep long groove commencing behind the ear and running along each side. This permits stretch for a big meal or a pregnant female. You will have to recognize several similiar species and many sub-species or simply call it a glass lizard.

Cnemidophorus sexlineatus - **six-lined racerunner.** To nine and one-half inches. From Maryland, south. A dull four-legged speed demon. Six light stripes and a long tapering slender tail.

Eumeces inexpectatus - **southeastern five-lined skink.** To eight and one-half inches, overall. From Virginia, south. Shiny, smooth, difficult to catch and also difficult to identify with certainty. Several similiar species.

SNAKES

No eyelids and no external ear openings. Three you may encounter are venomous. Let's review these poisonous species first.

Micrurus fulvius - **coral snake.** To four feet but usually only two feet. From North Carolina, south. Several species and subspecies. Rings of black, yellow, red, yellow and black in that order. *Yellow and red rings touch each other.* In the non-venomous ringed snakes, the red and yellow rings are separated by black. "Red to black, venom lack. Red to yellow, kill a fellow." If in doubt, do not approach it.

Agkistrodon piscivorous - **cottonmouth, water moccasin.** To four feet. From Virginia, south. Frequents water, looks much like *A. contortrix.* If in doubt, do not approach.

Agkistrodon contortrix - **copperhead.** To three feet. Color is variable in pattern of browns or reddish browns. It looks somewhat like a non-venomous milksnake, watersnake or hog-nosed snake, but has deep facial pits. Several subspecies from Massachusetts through Georgia. If in doubt, do not approach it.

The following snakes are non-venomous:*

Coluber constrictor - **blacksnake, northern black racer.** Five or six feet. From Nova Scotia, south. Several subspecies. Perhaps a little white on chin or throat, otherwise black all over. Slender, fast.

Heterodon platyrhynos - **hognose snake.** To about 30 inches. From New Hampshire, south. Looks like a venomous viper, hissing and puffing and coiling but traditionally considered harmless. It eats toads. I have seen this species swimming over oysterbeds enroute to tidemarsh islands. Nondescript brown snake with upturned nose.

Farancia erythrogramma, Abastor erythrogrammus - **rainbow snake.** To five feet. From Maryland, south. Glossy, irridescent, with red and black stripes. This snake eats eels, Yes, really! This snake really eats eels.

Farancia abacura - **mudsnake.** Up to five feet. From Virginia, south. Black back with pink pattern on belly. The tail ends with a really sharp point.

Regina rigida, Natrix rigida - **glossy water snake.** To two feet. From Virginia, south. Olive brown or brown with dark stripes. Two rows of black spots on belly. A really shiny watersnake.

Natrix sipedon - **water snake.** Several similar species. To 51 inches. From Maine, south. Variable. Ubiquitous. Eats fish and frogs. To be sure of the species, really sure, marry a herpetologist.

Natrix taxispilota - **brown water snake.** To five feet. From Virginia, south. It climbs, swims and fights with speed and vigor. It looks like a cottonmouth.

Opheodrys aestivus - **rough green snake.** To three feet. From New Jersey, south. Slender. Pale green. Sometimes semi-aquatic. Diet is mainly insects.

Thamnopis sauritus - **ribbon snake.** To three feet. Nova

Scotia to Georgia. A slim, agile, graceful garter snake. Several species. Semi-aquatic.

Thamnopis sirtalis - **garter snake.** Usually to two feet and rarely four feet. Entire Atlantic Coast. Yellow stripes and remainder of back is brown. Eats fish, frogs and earthworms. Many similar species. Less aquatic than ribbon snakes.

So then, Atlantic *Spartina* tidemarsh or on its edges we may find any of sixteen species of snakes, three of which are typically fanged and poisonous. A fly in the tidy ointment of our compartmented thinking comes from Dr. Sherman A. Minton, Jr., a physician and a highly respected herpetologist. He is the author of an article in Natural History Magazine, November 1978, in which he points out that many of our fangless traditionally non-venomous snakes including the garter and the hognose are in fact, demonstrably poisonous. As poisonous as a ratter? Well, no. How poisonous? Well, it varies.

It seems as if every time we get a good grip on a rule to box in Mother Nature, some iconoclast comes along and demolishes the box.

AVES, THE BIRDS

Many are called but few are chosen. The tidal marshes play host to tremendous numbers of many species of migratory birds, but not many are genuine resident breeders. There is no more point to list all the swallows and other casual visitors to the marshes than there is to list all the visitors to the Statue of Liberty. Please bear in mind that there are no trees on tidal marshes from Georgia north so strictly speaking we are limited to birds that nest on the ground, in reeds and cattails or in the low marsh elders. These elders are found at the highest elevation which can still be called a tidemarsh.

RAILS

Where to draw the line, where to say this is or is not a tidemarsh life form? It is easy when we consider the fiddler crabs, the ribbed mussel, the mummychug killifish, and the *Spartina* grasses—they thrive nowhere else, but it's not so easy with the mammals and birds. The typical mammals must surely be the otter and the muskrat, even though both are also found in nearly every other part of the U.S. regardless of salinity. The typical bird is easier to box in since it is locked into the tidemarsh almost as tightly as the fiddler crab; and of course it is the clapper rail *Rallus longirostris*. The clapper rail is *the* genuine salt water marsh hen from Georgia to Maine. Only in Florida is it found also in non-saline places. It resembles the fresh water marsh hen (*Rallus elegans elegans,* also called the king rail) so closely at times that the field identification should take into account the salinity of the nearest water. And the water will surely be near. The clapper rail even swims, which is unusual, considering that its feet are not webbed.

The clapper rail nests right on the tidemarsh, close to water; at high tide the base of the nest is frequently dampened. The young are, in the jargon, precocious—as soon as they are out of their shells and dried off, they are ready to travel through the saltmarsh grasses with the hen. The food of this rail is mostly animal and the typical tidemarsh decapods, isopods and amphipods are high on its list. Fiddler crabs and grass shrimp make up a large part of this rail's preferred diet and other crustacea, mollusks, aquatic insects, clamworms and small fish make up most of the remainder. What little plant matter they consume is also mostly tidemarsh in origin and it consists of cordgrass, bullrush, and sedge. Some intake of smartweed, acorns and soybeans has been recorded but from what we know of the habits of the clapper rail, these foods are not the first choices.

The king rail, Virginia rail, yellow rail, black rail and sora are similar species and their ranges overlap somewhat. Most casual visitors to tidemarshes will not see or even hear any rails on their first trip unless they are accompanied by a guide or a dog.

Paul Spitzer, a professional ornithologist, wrote an article about tidemarsh birds for a pamphlet published by the Conservation Commission of Old Lyme, Connecticut in 1977. With his permission I quote four paragraphs from that article because I believe he said it best:

"The casual observer never realizes that rails exist. They are wary birds that seldom leave the cover of dense marsh vegetation and they escape from men by running through cover, rather than flying. Their plumage is streaked with brown, olive-drab and other colors that blend with the marsh.

"The Connecticut River marshes are fairly crawling with rails ... a remarkable situation. What are you going to do about that? You can certainly hear a rail chorus, and maybe even see one. Dawn, dusk and darkness are the best times ... rails are active then. You can take a passive approach and go listen beside a quiet marsh, or you can take the proper calls from "The Field Guide to Bird Songs," Roger Tory Peterson's record, and replay them loudly on a cassette recorder. The rails think you're trying to stake out a territory, and the results can be surprising. These normally secretive birds may stride right up out of the marsh and try to stare you down.

"The grassy *Spartina* marshes in the lowest two miles of river support clapper rails, olive colored birds the size of a bantam hen, which make a clacking "cha-cha-cha" call, especially between April and September. Above this, where there is cattail, the smaller Virginia rail is common. Virginia rails can be summoned up even in the depths of winter. They make a descending "wunk-wunk-wunk" grunt which increases in pace. Unforgettable. Courtship takes place on nights in April and May and then the marsh resounds with a call that sounds like "tick-tick-tick-tick." This is known as the Hot Ticket call.

"When you hear or see a rail, you are becoming a naturalist. These birds don't appear all at once but the gradual discovery is central to the experience."

Sora

Clapper Rail

Marsh Hawk (Female)

MARSH HAWK

The marsh hawk is smaller than the osprey and it spends more time flying over marshlands than over marsh waters. Technically it is a harrier, *Circus cyaneus hudsonius*, formerly *Circus hudsonius*. When you are on a tidemarsh and you see its white rump patch as it rolls in flight, you have seen an excellent field mark. The tail is barred. Adult females are brown streaked and larger than the males. Adult males are solid pale gray and not streaked. But for their form and habits, one would not know the sexes were related. Tails are square and wingtips are dark. The wings are relatively narrow.

Marsh hawks are distributed all over the U.S., but they do breed on tide marshes and there is hardly a tidemarsh without a few. They eat all sorts of living prey including mice, muskrats, hares, rats, songbirds, ducks and barnyard poultry, reptiles, frogs and even insects.

Tidemarsh nests are often only inches higher than the high water mark and the eggs, about five, are oval white or pale blue-white. Some eggs have scattered pale brown markings on them. Both parents incubate the eggs beginning when the first is laid. They hatch in three or four weeks and the young fly a little only a month after hatching.

The marsh hawk is not a specialized eater. In one account 124 stomachs were examined and eight were empty. Of the remaining 116, 41 contained birds

79	"	mammals
7	"	reptiles
2	"	frogs
14	"	insects
143		

The discrepancy in addition is simply due to those stomachs which contained more than one category of food. Nearly all observers agree that meadow mice or meadow voles make up the major part of the marsh hawk's diet. Marsh hawks remain on most tidemarshes most of the year. In New England some brids may leave some marshes for the worst month or two, depending on the weather and perhaps also on the availability of mice.

BALD EAGLE

Actually, there seem to be two subspecies of *Haliaetus leucocephalus*. One is commonly called the northern and the other the southern bald eagle. Bald eagles are found almost all over the lower forty-eight states, Canada and Alaska, and I mention them briefly here because they frequently find their food in the tidemarshes and tidemarsh waters.

Adults are distinctive with their seven and a half foot wingspread, white heads (*leuco* = white, *cephalus* = head) and white tails. Juveniles are dark all over, and they resemble golden eagles for their first two or three years.

Bald eagles usually build tremendous nests in tall trees near water and so although they are not strictly marsh birds, they will be seen by visitors to the marshes. Both parents incubate two white eggs for about 35 days and the young remain in the nest for an additional ten or eleven weeks.

They eat *all* manner of animal food from rotten fish and carrion to rabbits, fish, coots and ducks. The bald eagle is a scavenger, a parasite on ospreys, and all the while a magnificent agile powerful predatory hunter thoroughly capable of capturing the most elusive game or swimming fish.

I have learned from experience as did the sagacious Benjamin Franklin that the bald eagle is much like motherhood and Social Security payments. No one should disparage either, but let's call a spade a spade. Those readers of this book who are irritated by my description but not familiar with A. C. Bent and his "Life Histories of North American Birds of Prey," Bulletin No. 170 of the Smithsonian Institution, United States National Museum should read that book before writing to me.

Strictly speaking, a parasite is one who eats at the table *of* another. This is not the same thing as one who eats at the same table *with* another. The latter situation is termed commensal. Some bird lovers would prefer to say that the bald eagle competes with the osprey for food; but since I never learned of an osprey obtaining a meal from an eagle, I will describe the eagle as an opportunistic or facultative parasite on the osprey. That is, the bald eagles steal fresh-caught fish from ospreys on a regular basis whenever the opportunity arises.

Eagles, ospreys and for that matter, all predatory birds are protected by United States law, and these laws are strictly enforced. Avoid visiting their nests and promptly report to the nearest conservation officer or game warden anyone who does.

OSPREY

The osprey is *the* fish hawk. It is unique among hawks in that its outside toe works equally well to the front or to the rear. Most hawks have three toes forward and one rear on each foot while the owls I've seen generally have two toes pointed each way. It is the only hawk that dines exclusively on live fish. It is found wherever there are fish, shores and temperate waters ... throughout the world and, like man, there is but one species. According to taxonomists there are five subspecies of osprey, but for ordinary bird-watchers, there is only one and nothing else quite like it. Taxonomists who watch ospreys are interesting to watch also: some of them are lumpers and some are splitters and most are so dogmatic that they might well form a separate race or even subspecies among *Homo sapiens* if for a few generations they should marry within their profession and then raise their children to be taxonomists.

Properly the international osprey is *Pandion haliaetus*, and our East Coast species is *P. h. carolinensis*. Like the muskrat it does very well on tidal marshes but it is also found inland wherever fish abound in open water. It breeds near water, feeds only from water, eats only live fish.

The general appearance of this hawk is graceful and slender. It looks blackish above and white below. Its head is distinctively marked with black cheek patches, and when the osprey soars, it holds its wings with a kink at the wrist joint. The female might stretch six feet from wingtip to wingtip and the male is generally a foot smaller.

This is a tremendous bird to be hovering and precision diving like a circus stuntman from a height of 100 feet into a rainbarrel, but that is just what the osprey does, and he

Osprey (Note one reversed claw)

does it very well indeed. So well that on occasion an over-size fish will not be raised nor will the bird release its grip and I am reminded of one of the closing scenes in Melville's "Moby Dick" where the harpooner Tashtego captures a sky hawk and pulls it down with him as the stricken Pequod sinks.

Ospreys nest near water, on rocks, on the tidemarsh ground, in tremendous tree nests which they build up over the years and on man-made towers, windmills, power poles, and cartwheels mounted on high masts. Americans often encourage ospreys to nest near their homes by providing these convenient nest sites. Perhaps this is a status symbol or perhaps a good luck omen akin to the stork's nest on the roof of a European home. The osprey nest also attracts other birds. Sparrows, swallows, martins and grackles frequently build their nests in the lower part of an elevated osprey nest. I speculate that small birds get protection by living close to the monarch, and perhaps they pick up scraps (or insects attracted by the scraps) or perhaps *they* do it for the status symbol.

When ospreys are north of Florida they are migratory. Of course they cannot feed through ice but there seems to be more to it than that. Ospreys seem to crave warmth. The migrators move north to Georgia in February, by March or April they are in Connecticut and by early June some hardy souls arrive in Alaska. The old nest is repaired and reoccupied or a new nest is built immediately and three multi-colored eggs comprise the usual clutch. The female incubates her eggs for 23 to 32 days and the male brings her fresh fish during this period. When he is not working, he generally perches nearby. After the eggs hatch the female shares the fishing effort but only she feeds the young. When they are about eight weeks old they learn to fly and soon thereafter they dive for fish. As the weather grows colder, ospreys move south; not in great flocks, but just a few at a time. They stop for a few days or a week at estuaries and lakes to rest and feed and then continue to keep ahead of the cold. Some go as far as the mainland of South America to winter, and some juveniles remain in the South for one or two or even three years before returning to the general area of their birthplace to mate and breed.

The osprey has had a lot of attention paid to it by environmentalists and ornithologists and egg collectors over the years. The eggs are certainly beautiful but today egg collections are a "no-no." The environmentalists are interested in ospreys because this bird is at the very end of a long food chain that often started on agricultural land. After the second World War the osprey population in New England began to lose ground at an accelerating rate. By 1960 it seemed that the handwriting was on the wall. The eggs broke or the embryos died in the shell. The cause was pretty well narrowed down to "PCB's" and persistent insecticides. The problem stemmed from the fact that industrial and pesticide poisons accumulated in short-lived animals in lesser and non-lethal concentrations and then built up to nearly lethal levels in the long-lived osprey. The chemicals inhibited the formation of strong shells and in some cases actually killed the embryo regardless of the strength of the shell. So the osprey became a symbol, not only of status, but also it become a symbol of our inability to live in harmony with our environment. Today, the use of persistent insecticides has been curtailed and the industrial poisons are regulated by law and chances are good that the osprey will remain with us. It should be clear to everyone that a bird that eats a few fish is absolutely meaningless in men's own ecosystem, but if they go, can we be far behind? Woops, this book is supposed to be a guide, but I got carried away again.

Ornithologists spend an inordinate amount of time with the osprey, partly because of its environmental problem and partly because it is such a large spectacular bird and partly because it lends itself to manipulation by humans. Ospreys are easy to capture. They can be banded easily and the bands are easy to see from a distance. Ospreys are easy to find, and easy to recognize. They live a long time. They return to the scenes of their youth. They certainly don't hide. Also, they tolerate a lot of funny business with their nesting arrangements. They will incubate extra eggs. They will incubate the eggs of others. They will feed fledglings taken from other nests. They are a delight for animal behaviorists; they even nest on cart wheels.

Short-eared Owl

SHORT-EARED OWL

Asio flammeus flammeus is the only owl known to nest on tidal marshes. Its most popular common name is "short-eared" but it is also known as the marsh owl, swamp owl and prairie owl. This is a large bird, with a wing spread of about 42 inches. Females tend to be larger than males and their colors are no good as fieldmarks because they vary tremendously between blotched gray and blotched brown.

As their name suggests, the ear tufts are not prominent and neither is the owlish facial disc. The tail is very short; on a perched bird it seems to be altogether missing. When it perches, two claws on each foot face forward and the other two point to the rear; this is generally the way with owls.

After a dramatic courtship with much high flying and wing clapping, it is almost a let-down when short-ears build their very simple nests right on the marsh. The nests are so low that sometimes the birds lose the clutch to the spring-perigee high tides. Other times, nesting pairs have been known to move their fledglings to higher ground to keep them from drowning. The eggs are plain buff-white and number four to seven. Both parents incubate the eggs (beginning with the first) and after about 21 days the eggs hatch in the order that they were laid. The young fly 31 to 36 days later and they hunt avidly during dawn, day and dusk as well as at night. They sleep for short intervals regardless of the time of day. And they eat—oh, how they eat! Their diet is mostly rodents, but once in a while an owl goes on a bird jag and sometimes picks a particular species to work over. This is not at all difficult for such a fast strong and acrobatic predator, but fortunately it happens only infrequently. Insects are also taken, but for a bird with a 42 inch wing spread, it would take a lot of insects to make a meal. Field mice, deer mice, house mice, meadow mice, voles, rabbits, small muskrats and even bats are the usual fare of this acrobat of the tidemarsh, the short-ear.

GULLS

There are many species of gulls, as you have probably noticed, but only three are frequent tidemarsh nesters. Typical of these is the ubiquitous herring gull, *Larus argentatus*. Like all the gulls in this family, *Laridae*, it is web-footed, large, coarse, sturdy, heavy billed and limited in adult plumage to white, gray and black. The specific name is derived from the Latin word for silver, *argentum*, and the overall appearance of the adults is silver-gray with some black on the tips of wing flight feathers. The young are dusky for their first two or three years, but by the time they are flying, they are the same size as their parents, from beak tip to tail tip, two feet long.

Herring Gull

Great Black Backed Gull

When the herring gull nests on tidemarshes, an infrequent occurrence to be sure, it builds a simple cushion with a central dent and the female lays three olive-green, brown blotched eggs. Both parents incubate the eggs and defend and tend the young, but if a fledging looks a little weak or stupid, they will probably eat it. Twenty eight years for a herring gull is probably the rule and not an exception. These birds are rough. They are found everywhere and they eat all manner of animal food, regardless of whether it is alive, dead, or long since dead.

The great black-backed gull *Larus marinus* is the big gull. His wings span five and a half feet and his length exceeds that of the herring gull by six inches. The name is enough to identify it in the field. The nest resembles that of its smaller cousin and so do the eggs. This big bird is a real terror at sea, on the garbage dump and in colonies of nesting sea birds. It has even been known to attack ewes and kill lambs. This bird is a genuine opportunistic predator who steals food from herring gulls and when all else fails, he scavenges.

The third gull of about eighty species known worldwide which actually nests on tidemarshes is the laughing gull *Larus atricilla*. It is tiny, when compared to the great black-backed, measuring only about 16 inches in length. Its head is black in summer and gray in winter, and its bill and feet are dusky red. The breast and underparts are white and the back is gray. Franklin's gull and the blackheaded gull resemble it. The eggs are small versions of herring gull eggs but the nest is sometimes under something of a bower of live grass, open at each end. The laughing gull calls *HaHa* so well that his name is a good fieldmark. By comparison with the aforementioned herring and great black-backed gulls, this little fellow is less of a garbage-dump scavenger and less of a predator upon other birds, but he is known to steal fish—literally from the beaks of pelicans. If I believed in anthropomorphism, I would equate his stealing of fish from the pelican with his laughter.

WILLET

This is but one of a tremendous number of slender-billed shore birds with long beaks, long legs and slender bodies that run along mud banks and half-dry creek bottoms at low tide. It is unusual on two counts. First, although it is not a swimming bird, it is semi-palmated (its toes are slightly webbed) and second, it is a genuine tidemarsh nester (most of the other similar species nest in sand dunes or prairies or tundras or obscure and romantic places).

Willet nests have been found on saltmarshes from Maine to Florida. These birds tend to colonize, and if you find one nest, the chances are good that there will be another 200 feet away. Four eggs with slightly glossy smooth shells are laid a day or two apart. The shell color varies from grayish to olive-buff with large darker blotches of brown. Incubation takes 22 to 29 days, with the female doing all the daytime setting. Perhaps the male helps out at night. Incubation commences after the first or second egg is laid and occasionally it is completed when the first or second chick hatches. This procedure sometimes leaves a few late embryos to die in their shells. Since the willet (more precisely *Catotrophorus semipalmatus*) nests so close to the highwater mark it may be that the seemingly wasteful incubation technique will produce some living young, even if there is flooding at either end of the nesting period.

The willet is about 15 inches long, with long legs, a relatively long straight bill, and no spectacular markings. It is a brownish gray bird with black and white wing patches. The tail is white and the legs blue-gray. Since no other slender billed shore bird is 15 inches long *and* gray *and* semipalmated *and* found actually nesting on tidemarshes, identification is easy even though its coloration is relatively nondescript.

The diet of the willet is ideal for a tidemarsh nester; fiddler crabs, mollusks, insects, small fish, roots and seeds such as rice.

I might properly mention again that precious few of the eighty or so species of sandpiper-like birds actually build their nests on our Atlantic Coast tidemarshes although many of them pass through and even remain awhile during their migrations. Some of these birds will travel over 10,000 miles a year from Arctic nesting ground to tropical paradise. Why? I wonder.

Willet

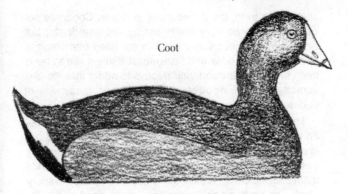

Coot

COOT

Fulica americana the coot has a name problem he probably isn't aware of. He has a plethora of common unscientific names and in addition, other birds are often called coots. To clarify the issue, first look at the bird, you will find it on tidemarshes from northern South America to southern Canada. It is green legged, graybodied and has a white bill. There is a white patch under the tail and when it flies the trailing edge of the wing shows a white border. The toes of the coot are not tied together with webbing, but each toe is furnished with a wide flap which surely helps it to swim. The coot is more nearly related to the rails than to the ducks, and on the marsh he does spend a good bit of time walking rather than swimming.

Back to names now, scoters are often called coots and this is confusing since scoters *(Melanitta* and *Oidemia)* are web-footed coastal sea ducks and not at all tidemarsh birds. Our tidemarsh coot has many local names. Perhaps you have known it as a crow duck, pond crow, sea crow, or blue peter, meadow hen, marsh hen, mud hen, moor hen, or pull-doo, white bill, American coot, ivory-billed coot, aplatter or shuffler. It, and the gallinules which follow were deliberately drawn in the style of wooden decoys. These birds do have a "wooden" look about them.

Coots act like rails as they push through *Spartina* stems. When they swim in ponds and tidemarsh creeks they act like ducks. In fact, they sometimes accompany ducks, and the ducks let the coots investigate a hunter's blind. If no shots issue forth, the ducks move in closer. Coots are not considered to be very intelligent by our standards, but since they are not usually good to eat (they often have a strong oily fishy odor and taste) most hunters like to have them swimming around their decoys to add a little life-like veracity to their wooden, cork or plastic facsimiles of ducks.

The coot nests in tidemarshes. The hen incubates a dozen eggs starting with the first one she lays. The eggs are speckled with brown or black on a buff or pinkish creamy base. As the chicks hatch out, the male collects and tends them until the clutch is all hatched out. The young are sooty black with orange hair-like feathers on the head and neck. The bill of the baby coot is red with a black tip.

Coots eat both plant and animal material including duck-hunters' leftover sandwiches, algae, rice, insects and clams. There are as many recipes for coot as it has names, but the few times I tasted it I wished I hadn't.

When a coot takes off from the water, he runs while working his wings and eventually usually makes it into the air for a short distance. Underwater, a coot will swim several hundred feet if pressed.

Purple Gallinule

GALLINULES

There are two species much like coots, both in their feeding and in their nesting habits. Like coots, gallinules pump their heads as they swim. Yankees who are not bird-watchers tend to expect coots in the North and gallinules in the South, but this is not so. Coots winter as far south as Florida, and gallinules breed or wander north every summer. Gallinules remind me of chickens that learned to swim poorly and learned to fly poorly and never learned to crow. Gallinules' toes are not webbed or even furnished with coot-type flaps.

The purple gallinule (*Porphyrula martinica*) is positively beautiful. The back is bronzy green and the remainder is a rich opulent shade of purple. The bill is red with a yellow tip, the legs are yellow and the forehead has a light blue shield.

The Florida gallinule (*Gallinula chloropus cachinnans*) lacks the color of his purple cousin. He more nearly resembles the coot. The field mark to watch for is a red bill on a gray bird. This species is also called the red-billed mud hen, water chicken and mud chicken.

Florida Gallinule

BLACK DUCK AND MALLARD

These are the best known and most common ducks of the tidemarsh. *Anas platyrhynchos,* the mallard is also likely the ancestor of the domestic white duck. Certainly they interbreed and certainly from time to time a hunter will see a white bird in a flock of mallards or a farmer will find a "green head" in his flock of domestic birds. These "green heads" also interbreed with black ducks and pintails.

Mallards generally nest on marshes and the female incubates eight to a dozen buff or nearly white eggs for about 26 days while the male absents himself to moult partially. This results in his "eclipse" plumage while she does all the work.

The distinguishing marks of the mallard are confusing because of the eclipse moults of the male, but the blue wing patches are always to be seen on both adults. The typical green head and neck of the adult male is the mark to look for when it is not in breeding-season eclipse. The birds grow to about three pounds and eat all manner of insects, crustacea, mollusks, plant shoots, grain and underwater plants. Although it is common on brackish tidemarshes, it is even more common in freshwater. The less spectacular black duck *Anas rubripes tristis* is actually a better example of a true tidewater marsh duck, especially on the Atlantic coast since here he is not likely to be found above the brackish water; with this one exception, the habits of these two birds are similar. The black duck is sometimes called the black mallard and the females of the two species are hard for man and duck alike to tell apart. Female mallards are somewhat lighter-colored than black ducks, but at 100 yards, one would be hard put to separate one from the other. For a long time black ducks with orange-red feet were considered still another race apart but more recently the informed opinion has it that the red-leg black duck is simply an older bird.

William Cullen Bryant had this to say about the black duck:

> *Vainly the fowlers eye*
> *Might mark thy distant flight to do thee wrong*
> *As, darkly lined upon the crimson sky,*
> *Thy figure floats along.*

Mallard (Drake)

Whistling Swan

Mute Swan

SWANS

There are four North American species. Two of them, the trumpeter and the whooper are never seen in the East. The other two, both weighing up to thirty pounds (and all-white as adults) are seen on Atlantic marshes. They are really big; remember that a large mallard rarely exceeds three pounds.

Whistlers, *Olor columbianus,* have black bills and occur in tidemarsh waters from Maryland to North Carolina. The other Atlantic swan is the European mute swan, *Cygnus olor.* It was imported to adorn estates and parks and now it is common in Maryland, New Jersey, on the north and south shores of Long Island and in the Hudson and various Connecticut estuaries. The mute swan may be a trifle larger than the whistling swan but more important for recognition is that the mute swan's bill sports a large black knob in front of the eyes and the remainder of the bill is orange.

When swans take off (or fly low) they seem to me to be just barely making it. Their primary wing feathers rub and squeak, and when one flies over me, I hear loud voices telling me to lower my head. Experts agree that swans are powerful and fast in flight, but for me in this instance, discretion seems the better part of valor.

Swans build large nests on the marsh just barely high enough to remain dry. The nest is sometimes lined with feathers appropriately called swansdown and four or five creamy white eggs are incubated in 35 or 40 days by the female. The young take several years to mature and the adults seem to live forever. Young swans are called cygnets, females are pens and males are cobs.

Swans are protected from hunters and they reciprocate by uprooting and destroying large areas of aquatic vegetation as a result of their voracious appetites. They also consume soft shelled snails and clams.

Many areas which formerly supported flocks of ducks are now out-of-bounds to these small birds. Additionally, smaller water fowl that swim too close to feeding swans have been killed for their audacity.

Brant

Canada Goose

CANADA GOOSE

There is nothing quite like it except the brant, and so identification is easy. A large white cheek piece behind the eye and a white rump, black neck and feet, brown-black wings and back, and paler breast make this big bird virtually unmistakable. And compared to any wildfowl except swans they are big. Big Canada geese, *Branta canadensis,* weigh ten pounds and span six feet. By comparison, the brant is smaller and lacks the white cheek-piece.

The nest is on the ground and unusually close to water. The female incubates her five or six creamy white eggs for a month and the male stands guard with a great deal of fidelity. The family group is the gander leading, then the greenish yellow young, followed by the goose. Ornithological tradition requires me to state that J. J. Audubon was attacked by a gander in Kentucky and he thought at the time that the bird broke his right arm. Geese mostly eat aquatic vegetation, and grain when they can find it.

To set a sometimes garbled record straight, Canada geese do not feed their young. They simply lead the precocious little ones to places where there is food, the adults may even point to edible things but the goslings feed themselves.

BRANT

This is the small "cousin" of the Canada Goose. It lacks the white cheek piece and has instead a small white mark on its neck beginning an inch or so behind the head.

Brant, *Branta bernicla,* fed almost exclusively on eel grass until a blight wiped out the East Coast eel grass. The birds that didn't starve adapted themselves to a diet of other aquatic vegetation including an algae called sea lettuce. This diet shift may be the cause for some conflicting views as to the palatability of brant. Prior to 1930 it was reported as an epicure's delight and more recently much less so. Now the eel grass is coming back and also hopefully the reputed delicious flavor of roast breast of brant. We are what we eat.

Their habits excepting diet are similar to those of the Canada goose.

SEASIDE AND SHARPTAILED SPARROWS

There are several species and subspecies of doubtful taxanomic status. A novice should consider himself lucky if he recognizes a seaside sparrow in the midst of other sharp-tailed sparrows. Or *vice-versa*. The genus is *Ammospiza*, and the species and subspecies are beyond the scope of this book. Seaside sparrows are often streaked on the breast and superficially resemble the song sparrow except that the latter sports a large central breast spot in addition to its breast streaks.

All these sparrows eat leaf hoppers, flies, bugs, insect larvae and other insects. Amphipods (sand fleas) are also an important part of their diet. Plant food includes seeds of cordgrass, wild rice and panicgrass.

Duck hunters sitting quietly in their blinds know by the rustling and chirping that they are surrounded by these little fellows, but rarely glimpse a whole bird except in flight. Audubon remarked in his journal in 1839 that "having shot a number of these birds, merely for practice, I had them made into a pie, which, however, could not be eaten, on account of its fishy savor."

When the seaside sparrow runs, it carries its body high and its tail pointed down. It nests in wetter portions of the marsh than does the sharp-tailed. Seaside sparrow nests are found in tussocks of black grass (*Juncus gerardi*), and even in cordgrass (*Spartina alterniflora*). The eggs number three to six and are speckled and blotched with brown on a pale green-white background. The female incubates the eggs for perhaps ten days and the male remains nearby and sings to her. The young are able to leave the nest in no more than nine days after they hatch.

MARSH WRENS

When the *Spartina* moves as though an animal moved it, but nothing shows, there is a good chance that a marsh wren was behind it. These little brown birds that can carry their tails straight up are so much a part of the marsh grass that they are taken for granted even by casual visitors. They look like they belong.

Wrens eat ants, flies, millipedes, amphipods, isopods,

beetles, grasshoppers, crickets, and caterpillars; even the remains of small vertebrate animals are commonly found in their stomachs.

There are between six and nine species of wren found in the United States and two are typical marsh inhabitants. These are the long-billed marsh wren, *Telmatodytes palustris,* and the short-billed marsh wren, *Cistothorus platensis stellaris.*

The long-billed is the larger of the two and it is more likely to be seen in cattails. Of course it is mostly brown, but a conspicuous white line over the eye accompanied by black and white stripes on the back, make it easy to identify. No other marsh wren has those black and white stripes. Don't start to measure bills—this will accomplish nothing.

The short-billed species is smaller and its back is striped with brown and white. There is no white streak above the eye but there is some streaking at the crown.

Wrens construct globe-shaped nests with side entrances from coarse grass with fine grass linings. The nests are attached to reeds or bushes. Chocolate brown eggs with olive or cinnamon markings are the product of the long-billed. White eggs or white eggs with lavender spots come from the short-billed species. Clutches consist of five to nine eggs.

Some wrens chatter incessantly and others make a sound that could come from striking a pebble against another pebble. Marsh hawks eat them.

REDWING

The male is jet black but for his shoulder marks of orange and scarlet. The black of the redwing blackbird, *Agelaius phoeniceus* spec. is the blackest of blacks and the red looks all the more red for the orange bar under it. There are several similar subspecies. These birds are larger than sparrows and smaller than robins but more important is the color. Cowbirds are colored muddy, so too are female redwings, but the male redwing has no look-alike or even look-similar.

Another thing about these year-round residents of the

southern marshes and summer visitors to the North is that the males arrive in New England two months before the females. This is said to be for the territorial staking of claims. Cattails and phragmites left over from the previous year become perches for male redwings who sing themselves hoarse for two months before the females arrive. Think of it—two months! There must be an easier way. The nest is woven into tall reeds and the female incubates about five blue eggs with their purple squiggley marks while the male sits nearby and continues to sing himself hoarse.

Redwing's diet is about 75% seeds and the remainder is insects. In some southern agricultural areas they are considered a pest.

BOAT-TAILED GRACKLE

Smaller than a crow, larger than a redwing, this yellow-or brown eyed large tailed, shiny, irridescent black bird is a real stunner. It is found from New Jersey, south in tide-marshes where it nests in marsh grass and low bushes. The genus is *Quiscalus,* formerly *Cassidix.*

The three or four eggs are pale blue with brown and purple spots and squiggles. Grackles are omniverous feeders and their diet ranges from rice to crustaceans and even small mammals.

BITTERN

The American bittern is a two or three foot bulky slow flying brown bird. You will flush it from all Atlantic *Spartina* marshes which are not iced over. During the cold months it is rare north of Virginia. It makes a number of interesting noises, mostly guttural, but sometimes duck-like and sometimes booming or braying.

There are no nesting colonies of bitterns but rather a single pair picks a territory and the female builds a sloppy nest of reeds on the ground. The eggs, three to six, are brownish and both the female and the male take turns to incubate them for about 26 days. Then the female feeds the young on regurgitated food for about eight weeks. This common bird is known scientifically as *Botaurus lentiginosus* and it was a favorite food of King Henry VIII of England. Bitterns eat fish, crustacea, insects, mice and smaller birds.

Redwing

You may also encounter a really tiny bird of bittern-heron shape, but no larger than a meadowlark. This little chap is difficult to flush and if you kick one out he will fly the bare minimum safe distance; then he seems to collapse back into the grass. This heron-shaped bird with a black back and buff wing patches is probably the eastern least bittern, *Ixobrychus exilis exilis*. It winters from south Georgia, south. In the mild seasons it will be found breeding as far north as Maine. The nest is a flimsy basket in cattails and the chalky white eggs number four to six.

American Bittern

GREEN HERON

This bird is not a tidemarsh nester but since it is so much a part of the tidemarsh when it is not reproducing itself, it deserves brief mention here. Green herons nest in trees. The nests are so sloppily made that you can frequently count the eggs by looking up through the bottom. Three to five greenish or blue-green eggs hatch in twenty days and on a diet of insects, shrimp, amphipods, crabs and a scattering of small fish, the young mature in six to eight weeks.

When you see a green heron on the tidemarsh you will first discover that it resembles a crow when it is flying and a bittern when it is hunting. It is not green. I should properly say I've never seen a green one but rather blue-green-gray with a pale grey belly and rusty throat.

The scientific name of the green heron is *Butorides virescens.* When it flies, its neck is tucked into its shoulders in an "S" curve. This is also true of the bittern, the egret, the great blue and the night herons, but not true of the ibis.

SNOWY EGRET

Egreta thula has an all-white plumage, black bill, yellow feet and black legs. This is the bird that was nearly made extinct by plumage hunters for the decoration of ladies' hats. It is usually a trifle larger than a green heron, looks more slender than the black-crowned night heron.

There are three or four pale blue eggs in a low nest hardly ever north of South Carolina. The nest may be in reeds or low bushes, often near nests of other species of herons.

NIGHT HERONS

There are two species and both sometimes nest in marsh reeds. Both are about the same size, approximately two feet from beak tip to tail tip. This makes them a little larger than the green heron and much smaller than the great blue. Their nests are sometimes in trees but regardless, the four or five eggs are pale blue-green. They winter where there is little or no ice.

The yellow-crowned night heron, *Nycticorax violacea,* is a dark gray bird with a yellow crown on its black head. The cheeks are white. You will find it from Massachusetts to

Yellow-crowned Night Heron

Black-crowned Night Heron

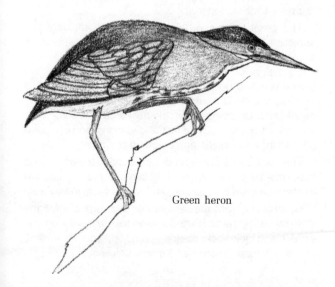

Green heron

Florida. It eats fish which it does not spear but pinches and it utters all sorts of odd squawks, and croaks.

The closely related black-crowned night heron *Nyctico-rax nycticorax* has a white face, a black back and a black crown. The tail is so short it hardly shows. This species is found from Maine to Florida in warm weather.

GREAT BLUE HERON

Not blue like a bluebird but rather blue-gray, he is so big and his neck and legs are so long, you would have trouble confusing him with any other bird on the tidemarsh. When he flies he kinks his neck back between his shoulders, his legs trail and seem to drag and his slow laborious wingbeat makes some people want to feel sorry for him. I don't. In 1938 one of them nearly blinded me with a jab of his long sharp rapier-like bill when I tried to rescue him during that great September hurricane.

Great blue herons *Ardea herodias,* are known to mate with great white herons, *Ardea occidentalis.* Perhaps the great white of Florida is just a color phase of the "great blue" found just about wherever there is water and mud with fish, crustacea, large insects like crickets and grass-hoppers, snakes, and frogs.

The great blue is certainly a tidemarsh resident and sometimes a tidemarsh nester. Although most nests are in trees within a short flight of tidemarshes, some pairs actual-ly build right on the ground. The three or four eggs are blue-green and both parents share in the incubation and feeding duties. The nests are sloppily built and sloppily maintained. Droppings paint the nest area and sometimes the smell alone suffices for fast field identification of a colo-ny, and they do tend to form breeding colonies.

There are perhaps five races not counting the whites and those that range farthest north in summer may be classed as migratory, but the southern individuals seem to stay put.

Great blue herons are tremendous, but surprisingly light. They stand four feet high and spread their wings six or sev-en feet, and their necks stretch several feet and with all that, they weigh hardly more than a spring chicken.

Great Blue Heron

IBIS

There are two species which nest colonially in trees but like the green heron spend a great deal of their time in our tidemarshes. They fly with neck outstretched which separates them from the heron. When you see one wading on clam mud the first thing you will notice is that its bill curves downward. In our range the dark bird is the eastern glossy ibis *Plegadis falcinellus falcinellus.* It is actually bronze colored but usually appears black from a distance.

The only other ibis you may see is the white ibis, *Guara alba,* of South Carolina, south. It is mostly white with black wingtips, a red face and red legs.

FORSTERS TERN

This bird looks like the common tern. Unless you are dedicated bird-watcher you will not be able to tell one from the other BUT *Sterna forsteri* (Forsters) sometimes makes its solitary nest on tidemarshes and the common tern most frequently nests in colonies on sand. Also, the eggs differ. Forsters' eggs are olive-brown with brown spots and lines, whereas the common tern lays pale brown or buff eggs with dark blotches.

As you visit tidemarshes you will surely see at least a half-dozen species of terns overhead but only Forsters is known to nest there. The greatest concentration of nests seems to be south of Maryland and it winters from South Carolina, south.

THE MAMMALS

Of course, mammals are the vertebrate animals that suckle their young. They have dorsally located central nervous systems, backbones, warm blood and hair. All female mammals nurse their young (hence their name). Even whales do it. A considerable variety visit tidal marshes from time to time, but very few species make it their permanent home. The marsh habitat is limited to its potentialities for nesting sites, food storage, hiding places; even the food supply is restricted in those areas where there is deep frost and much snow.

MUSKRAT

The mammal which is typical of both fresh water marsh and brackish tidal marsh life from the Gulf of St. Lawrence through Virginia and into North Carolina and also in the Mississippi, Louisiana and Texas Gulf coasts is, of course, the muskrat, *Ondatra zibethica,* also sometimes spelled *O. zibethicus.* An older scientific name is *Fiber zibethicus.* The common names include "rat" (a misnomer), musquash, and mud cat. There are perhaps a few subspecies, and these are of little interest except to mammologists.

Adults range in weight from two to four pounds. Total maximum length is twenty-five inches—the head and body making up fourteen. Females tend to be smaller than males. The tail is naked, black and flattened from side to side. It is the only North American mammal with a tail like *that.* Both sexes have glands which produce a musky odor.

The fur is rich and dense and overlaid with coarse guard hairs. When these guard hairs are removed and the pelt is dyed a little, it bears a striking resemblance to mink; and many fine garments bear the label "MINK dyed muskrat". Sometimes it has been called "Hudson Seal" or "River Mink." The fur whether natural or dyed is attractive, warm and durable. Many American boys have earned their nest-egg for college by trapping muskrats during the last two or three years of high school. A respectable trapline can be tended daily between school closing and suppertime. In December a flashlight glimmering on a tide marsh is frequently a young trapper who also squeezed in some basketball practice. Trapping muskrats is an industry which

Muskrat

doesn't get much attention from anyone who isn't directly involved. Trappers move quietly, following the creeks on foot or more likely in a rowboat pausing for a moment at an inconspicuous stick poked in the mud at or near the creek bank. The traps are usually not baited. They are placed underwater, strategically near but not on burrows and muskrat houses. Fur buyers take skins or fresh whole carcasses, depending on local custom. During the winter of 1976-1977 the best price for large prime skins paid to a trapper ranged from $5.00 to $6.50. Ten million pelts a year is a reasonable estimate of the nationwide industry.

In many ways the muskrat can be likened to a small beaver. It too is an amphibious rodent with a desirable pelt, but it adapts itself to more environmental niches. Ten million pelts a year suggests it is doing something right. For instance, it seems indifferent to salinity. What it drinks has always been something of a puzzle to me, but to get back, its food is generally stems, rootstocks and leaves of marsh plants, tidal or otherwise. Muskrats are also known to eat both water lilies and corn. Their diet also includes crabs, mussels, insects, and even fish when they can catch them.

The muskrat is a "family" animal. The conical mud and reed house occupied by a family rises out of the shallow water perhaps two feet above high water. There are no openings exposed to the air. All entry is accomplished by swimming. The house is a rugged affair. If you find an abandoned one and tear it apart you may not only find yourself in trouble with the law, but you will also find the going quite tough. The grass seems to be woven and cemented. Some houses on tidal marshes are six feet in diameter. Muskrat houses are occupied mostly in wintertime while burrows in creek banks lead to dens which are used during the remainder of the year. The muskrat does not hibernate. When he is not eating or building a house, the muskrat must certainly be busy reproducing itself, because even with all the trapping and predators, muskrats seem as plentiful now as they ever were. The average litter is seven but twelve youngsters are not rare, and two or three litters per year seem to be the rule.

Another muskrat name is marsh hare or marsh rabbit. It was not mentioned in the earlier paragraphs because these names refer only to the edible portion of the skinned carcass. A fat muskrat carcass is good to eat. It tastes like terrapin or wild duck, according to one authority. Apparently there is something amiss with our taste buds or my arithmetic since ten million muskrat carcasses with one pound of edible meat on each would represent a good part of our national meat diet.

Muskrats were brought to England in 1927 for the fur trade, but some escaped and did quite well undermining dams and levees and eating carrots, potatoes and corn to the point where the government tried to exterminate them—at no doubt considerable expense—all this in the course of ten years.

In places where both occur, alligators eat wild muskrats and also their larger South American counterpart, the nutria.

NUTRIA - COYPU

This is an exotic vegetarian marsh rodent, South American in origin, which was introduced as a fur producing animal for "ranch" propogation. The nutria, *Myocastor coypus* is most numerous in Oregon and Louisiana marshes; but appears elsewhere, Chesapeake Bay for example. It is large—up to 18 pounds with a body 25 inches long and a 17-inch tail. The tail is circular in cross section whereas the muskrat's tail is flattened from side to side and of course the beaver's tail is flat from top to bottom.

SMALL RODENTS AND
SIMILAR LOOKING CREATURES

The world abounds in rats and mice and moles and voles and shrews and lemmings. They are abundant, retiring, voracious and hard to tell apart. The moles and shrews are insectivores and not rodents at all. They eat animal food exclusively, whereas some small rodents are vegetarian and others are omniverous.

Shrews are difficult to identify. They all have beadlike eyes, tiny ears, five toes on each foot and soft fur. I suggest

Nutria

Meadow Mouse

you go directly to *A Field Guide to the Mammals* by Burt and Grossenheider to sort them out.

Moles are easier because there are fewer species and their tremendous forepaws and lack of external ears and virtual lack of eyes set them apart from other creatures but, to find out which one you have, you must again go to Burt and Grossenheider.

Rats have long tails and are larger than mice, many of which also have long tails. Both mice and rats have large prominent ears and eyes and only four toes on each front foot.

Voles tend to have relatively small ears (compared to our familiar house mouse *Mus musculus*) and compact bodies.

Woodrats have hairy tails (not scaly as on the Norway and Black rats) and soft fine fur. Their feet are white and so also are their bellies.

Rice rats and cotton rats are larger than mice and are found from New Jersey south and require a detailed examination by an expert for positive identification.

Lemmings look much like voles and at least one species, the southern bog lemming *Synaptomys cooperi* is found from South Carolina north.

I will proceed with these mammals with no apology but only after restating the going-on position. "This book is intended as annotated field guide for casual visitors." Only the prominent forms will be mentioned.

MEADOW MOUSE

From Georgia, north this seven-inch rodent is found on virtually every tidemarsh. It is, in point of numbers, and in consumption of vegetation and in providing food to predators like hawks and mink and foxes and bitterns and even gulls, a *most important creature.*

The meadow mouse *Microtus pennsylvanicus* is also known more properly but less commonly as the meadow vole. Many non-scientists call it the field mouse. It has a large stocky body and quarter pounders are not unusual. The tail is but an inch and a half long. The ears are small and the fur is relatively long. This rodent is a good swimmer; its fur is nearly waterproof, colored dark brown or gray and furnished with a few coarse guard hairs. Its eyes are dark, tiny, beady.

It is prolific. Prolific is an understatement. How about 7 + litters a year of four to eight youngsters in each litter?

The meadow mouse is active all year. It eats its own weight in food—up to a quarter of a pound—every day. According to L. L. Rue, the average density of this species is 30 per acre, but there is a record of 12,000 meadow mice per acre in one instance. Its diet is practically 100% plant food. It eats stems, leaves, flowers, seeds, roots, bulbs, bark. It hoards food in underground burrows against harsh winters. It is active in the snow and you will find vole runs all over upper tidemarsh meadows.

I mentioned previously that a tidemarsh might produce 5 tons of vegetation per year. Now just thirty meadow mice weighing a quarter pound each and eating their own weight daily for a year consume (30 mice x 0.25 pounds per mouse x 365 days = 2737.5 pounds per year) 1.36 tons out of the five tons produced; that's a lot of production destined for just one single consumer species.

WHITE FOOTED MOUSE

Here is another seven-inch rodent, the same length as the meadow mouse but only one-quarter the weight. *Peromyscus leucopus* is named for his white feet *leuco* = white, *pus* = feet. By contrast with the meadow mouse this species is dainty, slender and mostly tail. Its feet and belly are white and other parts are grey and/or brown, depending on its age.

It eats not only vegetable matter but also insects and slugs and snails. It is nocturnal, it lives in brushy places—for example under marsh elder shrubs at the upper edge of tidemarshes from North Carolina to southern Maine, and it spends a good deal of its time grooming itself. The white footed mouse is less a resident of tidemarshes than the meadow mouse, but you will surely see them on the upper meadows and shrubby edges.

COTTONTAIL

Most people lump the rabbits and hares with the rodents, but this is technically wrong. Look at their teeth. Rodents have two ever-growing front incisors in the upper jaw and a matching pair in the lower jaw. This leads to the name *simplicidentata* which you may see in taxonomic texts. Hares and rabbits have a second pair of smaller incisors behind the prominent pair in the upper jaw. This sets them apart, if the short cottony tail does not. For them you will see the name *duplicidentata* in the texts. All right, everyone knows the eastern cottontail *Sylvilagus floridanus,* common from Massachusetts, south. It is our most common, sought after, U. S. game animal. Thirty percent of the hunting ammunition purchased in the U. S. is expended on cottontail rabbits, but their numbers do not seem to suffer. The annual U. S. hunting harvest is about 25 million. Apparently it is habitat and not hunting that determines the rabbit population. Two additional species also appear in our range. The New England cottontail from New Jersey north to southern Maine and the marsh rabbit from Virginia, south. Call them all cottontails. They are abundant on the upper edges of tidemarshes where they squat under bushes and eat green stuff when they can find it; bark and twigs in northern winters.

Rabbits are promiscuous and prolific. They live in the open or in burrows made by other animals such as groundhogs. Young are born naked and blind, usually in litters of five to seven as frequently as four or five times a year. Incidentally, by contrast, the hare is precocious, its young are able to fend for themselves only a few hours after being born.

Maximum weight of a cottontail is three pounds for an eight-year-old buck. A really big doe might be a half-pound lighter.

When cottontails are not cut down by predators, they either exhaust their food supply or transmit parasites and disease to each other by crowding and then Mother Nature thins them out. Rabbits frequently suffer from tularemia, transmitted between animals by blood-sucking insects and then transmitted to man when he handles or eats an infected rabbit. A rabbit suffering from tularemia appears to be in a stupor. Kill, but don't touch it. Fortunately, thorough cooking utterly destroys the disease. Experienced hunters always handle or eviscerate any rabbit with gloved hands. Tidemarsh borders will support as many as four cottontails per acre.

PREDATORY MAMMALS

The predatory mammals of the tidemarsh usually den in dry hollow trees or sandbanks and by contrast with the muskrat, most don't spend their entire lives on the wetland. In the same breath I must say that no tidemarsh I ever saw or heard of was not routinely visited by a flesh-eating mammal.

By way of a quick review I will list those you may see or hear or find the tracks of in your travels.

Tabby, the domestic **cat** *Felis domesticus* will frequently appear on the upper slopes of tidemarshes near human habitation. Object: mice.

Fido, the domestic **dog** *Canis familiaris* will also visit tidemarshes and will wreak havoc with nesting ducks if not controlled. Dogs in packs will drive deer into bogs or creeks when the opportunity is presented.

Reynard, the **fox** may be the red *Vulpes fulva* or the gray *Urocyon cinereoargenteus.* These are primarily mousers but will also take a cottontail or a nesting bird if the opportunity arises.

Ringtail, the **raccoon,** *Procyon lotor* is another frequent visitor to tidemarshes. In northern climates, if the winter is fierce, this animal will den up for a while but where there is food year-round, he is active year-round. The 'coon is fairly close to being a full-time resident of tidemarshes. He eats virtually every creature larger than a pinhead that is mentioned in this book and some of the plantlife as well. He thrives on marshes and will sometimes reach a weight of forty pounds. He swims, he creeps, he burrows, he digs and pries and probes. The coon is an inveterate investigator. You will find his tracks and scats everywhere, don't be surprised.

River Otter

The **otter,** *Lutro canadensis,* a close relative of the weasels, mink, marten, fisher, wolverine, ferret, badger and skunk, is as nearly the 100% genuine tidemarsh large predatory mammal as we are likely to find. It is smart, so smart that domesticated pet otters have been trained to retrieve ducks for hunters. They live in wet places from Maine to Florida and every other of the 49 continental states and Canada as well. This is properly the river otter and males grow to weigh as much as 25 pounds. Females are 25% smaller. The otter builds its den in a bank, often with the entrance under water.

It is sociable, smart, long-lived (over 14 years), valuable for its pelt, and most tidemarsh visitors will never get to see it simply because it is elusive and intelligent.

The **mink**, *Mustela vison* derives its name either from *Mus* as the musky one like the mouse or the seeker of mice. In Sanskrit "mus" suggests "thief". No matter, it is a musky smelling aggressive predator of all marshes including tidemarshes. It eats rodents, birds, eggs, fish, frogs, crustacea. Ladies revel in the luxury of its pelt, this mink. Today relatively few mink are trapped for the fur trade, most are raised in cages on ranches. If you ever want to get a little more motion from a lazy horse, suggest that his next trip might well be to the mink farm. Mink are mostly nocturnal and when they are active, they are very active. They are great swimmers even though their feet are not webbed as are the otters'. The mink is intermediate in size between the chipmunk-sized weasel and the big cat-sized otter. A large male might measure 17 inches overall and weigh three pounds. A female would be somewhat smaller.

Wood pussy, the **skunk**, *Mephitis mephitis,* is still another predator of the tidemarsh. It is slow, deliberate, persistent and bold. Few skunks retreat from anything less formidable than the great horned owl. One whiff will tell you why. He is completely omniverous and although he retires in inclement weather, he does not hibernate. If things outside are really rough he might den up and simply sleep it out for a few days. A young skunk is easy to tame and makes an interesting pet. Old wild skunks are best avoided, especially if rabies has been reported, since they have been known to transmit it. Skunks are frequent visitors to tidemarshes but since they are nocturnal and the tides tend to wash away the tracks, you may never meet one. Just as well.

The **opossum**, *Didelphis marsupialis,* may grow to thirteen fat ugly gray pounds. He is nocturnal, very strong, very stupid, fairly slow and usually hungry. His rat-like tail resembles that of the muskrat or nutria, but you will never mistake it for anything but a 'possum. If in doubt, look at the teeth. Muskrats and nutria have the gnawing incisors of rodents, but the 'possum has a headful of dog-like teeth including prominent canines. Opossums scavenge and search out birds' nests and all manner of fruits and vegetables. This marsupial, once known as the Virginia opossum is a slow but strong swimmer, so don't be surprised to find it foraging on tidemarsh islands from Massachusetts, south.

WHITETAIL DEER

Someday a reader of this book will be canoeing through a *Spartina* or cattail lined creek and will hear a tremendous struggling crashing splashing pandemonium ahead, out of sight, around the next bend. Poor reader; you will wonder whether you are about to rescue a maiden in distress or be gobbled up by the Loch Ness monster. Actually, what more likely happened was that you startled Bambi, the whitetail deer *Odocuileus virginianus* and he was just leaving. Males grow to a record 400 pounds, live weight, and females reach 250 pounds, but these are the records, most adults are but half that. Deer are not mute but will occasionally let off a loud whistling snort if they sense your presence. These reed browsers are often seen on or near tidemarshes from Maine to Florida.

MANATEE

Among those few mammals that made it in *tidewaters* is the manatee (*Trichechus manatus*) from Florida to North Carolina. This is the sluggish mermaid with front flippers and fat lips. It grows to thirteen feet and is unforgettable once seen. It is protected everywhere. Manatees are 100% vegetarian and 100% aquatic. They favor shallow lagoons and rivermouths. Since they enjoy eating that pest water weed, the water hyacinth, they provide an important service in opening up waterways for circulation. Unfortunately, they are sometimes injured or killed in collisions with motorboats.

CHIROPTERA

When you visit a tidemarsh on a warm still night you should find the air alive with animal movement, indistinct, fast, dark-colored in a dark sky. Even if you don't recognize them by their squeaks you may suspect that these creatures are not insects or birds if only by their movements. They fly rapidly, ever twisting, flitting is a good word. These are bats. They are nocturnal insect eaters and although they do not sleep on the tidemarsh, some do spend all their active hours flying over them.

THE GENERAL DISCLAIMER

Were this a boxtop, it might state "If you use this product and death results, you have no one to blame but yourself." Since this is a book about natural history, the disclaimer is directed instead toward Mother Nature and Her Works.

Every sailor and every hymn singer remembers the stirring lines that commence: "Eternal Father strong to save whose arm doth bind the restless wave, who bids't the mighty ocean deep its own appointed limits keep." Well this may be great on the high seas, but on the tidemarsh, the limits are indistinct and that is a challenge to any tidemarsh guide, man or book.

The animals and plants I have described are successful because they are flexibly adjusted to their variable and unpredictable environment. They make accommodations—to drink more or less salty water and still maintain optimum body fluid levels, to compensate for water level changes, to weather storms. They survive and struggle and press and compete. They push their appointed limits.

All this while Anglo-Saxon laws and Anglo-Saxon language tend to make us favor sharp boundaries and precise descriptions.

Alas, on the tidemarshes there are no absolutely appointed limits, no perfectly sharp boundaries, no precise definitions. Here are no pure blacks or pure whites, but rather many subtle shades of grey. For example, I know I wrote that poison ivy does not grow on the tidemarsh, but I also know of a place where a clump of poison ivy seemingly springs from a bed of typical tidemarsh *Spartinas* and sedges. The creek immediately adjacent supports blue crabs and barnacles. Red-jointed fiddler crabs burrow among the stems. Perhaps someday you will find a stand of wild rice growing alongside an oyster bed.

Well, now you have my boxtop disclaimer and furthermore I'm still itching from that poison ivy.

SELECTED REFERENCES

This little book was written, illustrated and designed to go with you onto a tidemarsh. Hopefully it will be your key to a wonderful experience in observation, identification and understanding.

The positive identification of similar species should start with the field guides and field books published by the Audubon Society, Putnam and Houghton-Mifflin and The Golden Press.

The ecology, botany, zoology, water chemistry and tidal forces are harder to find in popular literature. Most of the accurate and detailed information is scattered through periodical scientific publications and a few scholarly texts. I referred to some as they related to particulars, but here the intent is to mention the best books I know about for each specialty.

Lower Animals and Seaweeds, Natural History — Arnold, Augusta F. 1901. "The Sea Beach at Ebb Tide". The Century Company. Republished by Dover 1968. 490 pages. Illustrated. Up to but not including amphioxus.

Lower Animals, Identification — Miner, Roy W. 1950. "Field Book of Seashore Life" G.P. Putnam's Sons. 888 pages. Illustrated. Up to and including amphioxus, the lancelet.

Marine Animals, Natural History — MacGinitie, G.E. and Nettie. 1968. "Natural History of Marine Animals". McGraw-Hill. 523 pages. Illustrated. Protozoa to whales with most emphasis on lower forms.

Fishes, Identification and Habits — Breder, Charles M. 1948. "Field Book of Marine Fishes of the Atlantic Coast (from Labrador to Texas)". G.P. Putnam's Sons. 333 pages. Illustrated. Identification and habits.

Fishes, Indentification and Habits	Bigelow, Henry B. and Schroeder, William C. 1953. "Fishes of the Gulf of Maine". Fishery Bulletin 74 of the Fish and Wildlife Service. Volume 53. Reprinted 1964. Museum of Comparative Zoology, Harvard University, Cambridge, Mass. 02138. 577 pages. Illustrated. Identification and habits.
Wildlife Food Habits and Plants	Martin, Alexander C., Zim, Herbert S. and Nelson, Arnold L. "American Wildlife and Plants (A Guide to Wildlife Food Habits)". 1951. McGraw-Hill. Republished by Dover 1961. Illustrated. Ranges and diets of birds and mammals.
Legal and Historical Perspectives	Carroll, John E. 1974. "Protection and Preservation of Coastal Saltmarsh on the Northeast Atlantic Coast". Michigan State University. Ann Arbor, Michigan. No. 74-19, 793. 268 pages.
Birds, Identification	Peterson, Roger T. 1963. "A Field Guide to the Birds". Houghton-Mifflin. Illustrated. 290 pages. Field marks, voices and ranges.
Birds, Natural History	Bent, Arthur C. "Life Histories of North American Birds". Dover. A series of books originally published by the Smithsonian Institution. U.S. National Museum.
Mammals, Identification	Burt, W.H. and Grossenheider, R.P. 1952. "A Field Guide to the Mammals". Houghton-Mifflin. Illustrated. 200 pages.
Turtles, Identification and Natural History	Carr, Archie. 1952. "Handbook of Turtles". Comstock. 542 pages. Illustrated.

Reptiles and Amphibians, Identification	Conant, Roger. 1958. "A Field Guide to Reptiles and Amphibians". Houghton-Mifflin. 366 pages. Illustrated.
Crustacea Natural History	Schmitt, W.L. 1965. "Crustaceans". University of Michigan Press, Ann Arbor, Michigan. The best introduction.
Crusteacea Key	Bousfield, R.D. 1974. "Shallow Water Gammaridean Amphipods of New England". Cornell University Press. Ithaca, New York. This is simply a key.
Invertebrates	Gosner, K.L. 1971. "Guide to Identification of Marine and Estuarine Invertebrates", Wiley-Interscience, New York. This is simply a key.
Invertebrates	Olmstead, N.C. and Fell, P.E. 1974 "Tidal Marsh Invertebrates of Connecticut". Connecticut Arboretum at Connecticut College, New London. 36 pages, illustrated. Especially useful for spiders and insects.
Tides	Defant, Albert. 1958. "Ebb and Flow". University of Michigan Press, Ann Arbor, Michigan.
Mollusks, Natural History and Identification	Rogers, Julia E. 1908. "The Shell Book," Doubleday, Page. 485 pages. Illustrated.
Insects, Identification	Lutz, Frank E. "Field Book of Insects." G.P. Putnam's Sons. Illustrated.

Insects, Natural History	Frost, S.W. "General Entymology," McGraw-Hill republished in 1942 as "Insect Life and Insect Natural History". 526 pages. Illustrated. Dover.
Plants	Roberts, Mervin F. 1969. "Tidal Marshes of Connecticut". (A primer about the plants that grow in our wetlands.) Connecticut College Arboretum. New London, Connecticut. 32 pages. Illustrated.
Plants	Knobel, Edward 1977. "Field Guide to the Grasses, Sedges and Rushes of the U.S." Dover, New York.

INDEX

NOTES

NOTES

NOTES

NOTES

NOTES

Hybrids of Striped Bass (page 149) and White Perch (page 152) have been described and the close resemblance of these two species has been officially recognized.

The scientific name of the striper has been changed to reflect this close relationship. Both species are now placed in the same genus. Since *Morone* is the older generic name, it has precedence. *Roccus* has been dropped and the familiar striped bass is now, scientifically, and more reasonably, *Morone saxatilis*.

Common reedgrass (page 48) has a new scientific name. It is no longer *Phragmites communis* but should be called *Phragmites australis*. The non-scientific name fragmites is also frequently spelled the same as the generic name, *i.e.* phragmites but when it is used that way (page 34) it is not printed in italics.

ABOUT THE BOOK

Tidemarshes are often overlooked and misunderstood frontiers. They look forbidding, monotonous, uninhabited and sticky. Most visitors arrive only by mistake and then try to leave by the fastest possible means. They miss a lot. Marshes are really beautiful, life is abundant and diverse. Travel is neither difficult nor dangerous if one knows how to go about it.

This book is your guide to finding and recognizing several hundred plants and animals in the water, mud and airspace of grassy tidemarshes from Nova Scotia to Georgia.

ABOUT THE AUTHOR

Mervin Roberts has solid scientific credentials but more important, he has spent a good part of his life digging, hunting, fishing, exploring and studying tidemarshes. His biography appears in American Men and Women of Science and Who's Who in the East.

The publisher and the ISBN number shown elsewhere in this book have been changed. The new ISBN number is 0-933614-19-5.

This book is now distributed by:

Peregrine Press
Old Saybrook
Connecticut 06475